T0212650

AESS Interdisciplinary Environmental Studies and Sciences Series

Series Editor

Wil Burns, Forum for Climate Engineering Assessment, School of International Service, American University, Washington, DC, USA

Environmental professionals and scholars need resources that can help them to resolve interdisciplinary issues intrinsic to environmental management, governance, and research. The AESS branded book series draws upon a range of disciplinary fields pertinent to addressing environmental issues, including the physical and biological sciences, social sciences, engineering, economics, sustainability planning, and public policy. The rising importance of the interdisciplinary approach is evident in the growth of interdisciplinary academic environmental programs, such Environmental Studies and Sciences (ES&S), and related 'sustainability studies.'

The growth of interdisciplinary environmental education and professions, however, has yet to be accompanied by the complementary development of a vigorous and relevant interdisciplinary environmental literature. This series addresses this by publishing books and monographs grounded in interdisciplinary approaches to issues. It supports teaching and experiential learning in ES&S and sustainability studies programs, as well as those engaged in professional environmental occupations in both public and private sectors.

The series is designed to foster development of publications with clear and creative integration of the physical and biological sciences with other disciplines in the quest to address serious environmental problems. We will seek to subject submitted manuscripts to rigorous peer review by academics and professionals who share our interdisciplinary perspectives. The series will also be managed by an Editorial board of national and internationally recognized environmental academics and practitioners from a broad array of environmentally relevant disciplines who also embrace an interdisciplinary orientation.

Anne Egelston

Worth Saving

International Diplomacy to Protect
the Environment

Anne Egelston
Department of Chemistry, Geoscience,
and physics
Tarleton State University
Stephenville, TX, USA

ISSN 2509-9787   ISSN 2509-9795 (electronic)
AESS Interdisciplinary Environmental Studies and Sciences Series
ISBN 978-3-031-06992-5   ISBN 978-3-031-06990-1 (eBook)
https://doi.org/10.1007/978-3-031-06990-1

This Springer imprint is published by the registered company Springer Nature Switzerland AG
The registered company address is: Gewerbestrasse 11, 6330 Cham, Switzerland

Acknowledgments

As I place the finished manuscript in Springer's more than competent hands, I am more cognizant than before how much I have benefitted from the talents, friendship, and loyalty of my family, students, and coworkers. Each of you has supported, inspired, and spurred me on to complete this work. First, I want to thank my beautiful parents, Thomas and Carol Egelston. They are a constant source of pride and admiration for me. They provide me with my own personal writer's retreat, complete with space to write, good company, and all the comforts of my other home. I also want to thank my grandfather, Wayne Lewallen, still with us in his 93rd year and still walking around campus at Tarleton State University in Stephenville with me. While writing my first book, he told me that having an honorable mention in an acknowledgment was all the college experience he would receive. I believe you have exceeded expectations.

I am especially grateful to my undergraduate students for possessing the courage to speak out about the strengths and weaknesses of my classes. This book is intended for you. I hope this book balances the academic rigors of a scholarly publication and the provision of knowledge of what my students should learn to be successful in this realm of international environmental affairs in the future. Thus, if I have departed from the scholarly tradition of a knowledgeable reader in the subject area, I did so for a good cause.

In laboring on behalf of my undergraduate students, I have been especially assisted by a group of graduate student research assistants: Clifford Curry, Rachel Pozzi, Wanda (Matison) Rhodes, A'dayr Shewmaker, and Kolin Yancey. Each of you brought your own gifts to this process and I hope you have learned as much from me as I have from you. Anna Ronck, an undergraduate student worker, joined this group at the last minute to assist with indexing. I look forward to seeing what you do next.

My colleagues have listened to me both celebrate and despair in the course of the last two years. I would be remiss without thanking Dr. Ryan Morgan for helping to create a happy work environment and Dr. Amy O'Dell for helping me with all the administrative work that inevitably comes with such a large undertaking.

Last, I want to thank my administration at Tarleton State University. Working through the COVID-19 pandemic has not been easy. You have nevertheless paid attention to the needs of the students, faculty, and staff. I especially appreciate the "banked hours" release for the Spring 2022 semester that allowed me to complete this work.

Contents

Abbreviations

2030 ASD	2030 Agenda for Sustainable Development
AGLTE	Ad Hoc Group of Legal and Technical Experts on Biological Diversity
AIDS	Acquired Immunodeficiency Syndrome
AOSIS	Alliance of Small Island States
Bamako Convention	Bamako Convention on the Ban of the Import into Africa and the Control of Transboundary Movement and Management of Hazardous Wastes within Africa
Basel Convention	Basel Convention on the Control of Transboundary Movements of Hazardous Wastes and Their Disposal
Basel Protocol	Basel Protocol on Liability and Compensation for Damage Resulting from Transboundary Movements of Hazardous Wastes and their Disposal
BASIC	Brazil, South Africa, India, and China
BAT	Best Available Techniques
BBC	British Broadcasting Company
BCSD	Business Council for Sustainable Development
BINGO	Business and Industry Non-governmental Organization
BRS	Basel, Rotterdam, Stockholm
Cartagena Protocol	Cartagena Protocol on Biosafety to the Convention on Biological Diversity
CBD	Convention on Biological Diversity
CERCLA	Comprehensive Environmental Response, Compensation, and Liability Act
CFCs	Chlorofluorocarbons
COP	Conference of the Parties
CSD	Commission on Sustainable Development
DDT	Dichlorodiphenyltrichloroethane
ECOSOC	Economic and Social Council
EEZ	Exclusive Economic Zone
ENB	Earth Negotiations Bulletin

EU	European Union
Expert Group	Ad Hoc Groups of Experts to the Executive Director of UNEP
FAO	Food and Agricultural Organization
G-77	Group of 77
GATT	General Agreement on Tariffs and Trade
GEF	Global Environment Facility
GEMS	Global Environmental Monitoring System
GMOs	Genetically Modified Organisms
HBFCs	Hydrobromofluorocarbons
HCFCs	Hydrochlorofluorocarbons
Helsinki Convention	Convention on the Protection of the Marine Environment of the Baltic Sea Area
HFCs	Hydrofluorocarbons
IAEA	International Atomic Energy Agency
IBRD	International Bank for Reconstruction and Development
ICJ	International Court of Justice
IFC	International Facilitating Committee
IFCS	International Forum on Chemical Safety
IIED	International Institute for Environment and Development
ILO	International Labor Organization
IMF	International Monetary Fund
INC	Intergovernmental Negotiating Committee
Infoterra	International Referral System for Sources of Environmental Information
INTERPOL	International Criminal Police Organization
IPCC	Intergovernmental Panel on Climate Change
IRPTC	International Register of Potentially Toxic Chemicals
IUCN	International Union for Conservation of Nature
IWC	International Whaling Commission
LDC	Least Developed Countries
LMOs	Living Modified Organisms
MAP	Mediterranean Action Plan
MARPOL	International Convention for the Prevention of Pollution from Ships
MDGs	Millennium Development Goals
MED POL	Coordinated Mediterranean Pollution Monitoring and Research Program
MFN	Most Favored Nation
MOP	Meeting of the Parties
Nagoya Protocol	Nagoya Protocol on Access to Genetic Resources and the Fair and Equitable Sharing of Benefits Arising from their Utilization to the Convention on Biological Diversity
NASA	National Aeronautics and Space Administration
NAZCA	Non-state Actor Zone for Climate Action

NGOs	Non-Governmental Organizations
NIABY	Not in anybody's backyard
NIMBY	Not in my backyard
OECD	Organization for Economic Co-Operation and Development
OWG	Open Working Group
PCBs	Polychlorinated Biphenyls
PCDDs	Polychlorinated Dibenzodioxins
PCDFs	Polychlorinated Dibenzofurans
PD4SDG	Partnership Data for Sustainable Development Goals
PIC	Prior Informed Consent
POP	Persistent Organic Pollutant
PrepCom	Preparatory Committee
RCRA	Resource Conservation and Recovery Act
Rio + 20	United Nations Conference on Sustainable Development
Rotterdam Convention	Rotterdam Convention on the Prior Informed Consent Procedure for Certain Hazardous Chemicals and Pesticides on International Trade
RSP	Regional Seas Programs
SDGs	Sustainable Development Goals
Stockholm POPs	Stockholm Convention on Persistent Organic Pollutants
TRIPS	Trade-Related Aspects of Intellectual Property Rights
TSCA	Toxic Substances Control Act
UN	United Nations
UNCED	United Nations Conference on Environment and Development
UNCHE	United Nations Conference on the Human Environment
UNCLOS	United Nations Conference on the Law of the Sea
UNCLOS II	Second United Nations Conference on the Law of the Sea
UNCLOS III	Third United Nations Conference on the Law of the Sea
UN DESA	United Nations Department of Economic and Social Affairs
UNDP	United Nations Development Program
UNEP	United Nations Environment Program
UNESCO	United Nations Educational, Scientific, and Cultural Organization
UNFCCC	United Nations Framework Convention on Climate Change
UN GA	United Nations General Assembly
UV	Ultraviolet
Vienna Convention	Vienna Convention for the Protection of the Ozone Layer
WCED	World Commission on Environment and Development
WHO	World Health Organization
WMO	World Meteorological Organization

Working Group on ABS	Ad Hoc Open-Ended Working Group on Access and Benefit Sharing
WSSD	World Summit on Sustainable Development
WTO	World Trade Organization
WWF	World Wildlife Fund

Chapter 1
Introduction to International Environmental Politics

Abstract International environmental diplomacy will celebrate the fiftieth anniversary of its first mega-conference, the United Nations Conference on the Human Environment in the summer of 2022. This book presents an overview of major conferences and events that form the backdrop for international environmental diplomacy with the objective of assessing the successes and failures of international environmental diplomacy as well as understanding how scholars analyze the field. Three key characteristics of international environmental diplomacy are introduced, including change, continuity, and complexity. Together, these three characteristics describe the interrelationships and interdependencies of international environmental diplomacy. While scholars tend to simplify international environmental diplomacy into regimes and negotiating sessions with concrete beginnings and endings, this approach disaggregates contemporary events.

Keywords International environmental diplomacy · Environmental problems · International state system · International relations · Environmental governance · Environmental symbols

In October 2019, Extinction Rebellion staged a series of protests known as the "autumn uprising" to demand that the United Kingdom take more dramatic action to reduce greenhouse gas emissions. These public protests have successfully disrupted individual and collective routines on behalf of our global environment. Individual events included throwing fake blood at the Treasury building in Westminster, London [7], activists chaining themselves to the Marshall bridge in Berlin [4], and blocking traffic in New York's Times Square [1]. The disruptive character of the Extinction Rebellion's London campaign prompted the Metropolitan Police to ban the protestors from further action in London. However, this ban was quickly ignored and then revoked by the High Court of Justice in London [15].

These protests have sharpened focus and attention on climate change, and this type of protest technique remains a viable tactic for environmental movements worldwide. Iconic imagery such as whales, polar bears, and the Amazon rain forest reminds people of the need to protect and conserve the environment. However, even more, bleak and sinister imagery portrays the consequences of the failure to control the excesses of technological innovation, from the not so benign aerosols that caused the

© The Author(s), under exclusive license to Springer Nature Switzerland AG 2022
A. Egelston, *Worth Saving*, AESS Interdisciplinary Environmental Studies
and Sciences Series, https://doi.org/10.1007/978-3-031-06990-1_1

ozone hole to the nuclear fallout signs that attempted to provide a safe space in case of nuclear war. More damning evidence of humankind damaging the environment came with the creation of the Chernobyl Exclusion Zone, an area roughly 2600 km^2 (1000 m^2) in size with limited public access due to nuclear radiation. Founded in 1986 after the Chernobyl nuclear reactor meltdown in Ukraine, the area is unfit for human settlement due to extremely high radiation. However, a macabre tourist industry based in the town of Chernobyl (outside of the Exclusion Zone) takes tourists inside the area for the day. Due to the uneven exposure of radioactive particles, some less impacted sections within the Chernobyl Exclusion Zone appear to be recovering. However, the area immediately adjacent to the nuclear plant remains uninhabitable.

These ghastly reminders of the consequences of environmental mismanagement also prick the conscience toward environmental protectionism. Regardless of the positive or negative imagery used to represent the health of our planet, the global public cares about environmental issues, and these images spur some people to engage in environmental advocacy at all levels of government—the local, national, and international levels. These pro-environmental voices seek changes in our society to improve the quality of the environment. These voices sometimes successfully force change by clamoring for new laws that prevent others from engaging in activities that pollute land, air, and water. In other instances, environmental activists promote activities that prevent environmental damage, such as utilizing refillable water bottles or purchasing printer paper with high recyclable fiber content. Alternatively, there are also instances where pro-environmental voices lose out to other concerns, most often economic concerns espoused by businesses, industries, and shareholders that profit from unfettered economic growth.

Environmental symbols, however, do not point solely to our current policy choices; they also remind us of past policy preferences already enacted. Why is the whale a prominent symbol of environmental affairs recognized worldwide? While the visual impact of the origins of the Merry Pranksters driving around Stockholm in 1972 with a papier-mâché whale on top of their much more memorable school bus, Furthur, has faded, the collective memory of Greenpeace's "Save the Whales" campaign has not. This campaign began in April 1972. By 2010 this campaign gained legendary status, with frequent reminders of one of the most poignant and prominent successes of the environmental movement enduring in contemporary pop culture. Scholars such as Kalland [20] suggest that the whale was emblemized by non-governmental organizations (NGOs) such as Greenpeace, who successfully reframed the conversation around whether to hunt whales into a good guys (whales)/bad guys (whalers) argument. Scholars and practitioners credit this campaign for the 1982 ban on commercial whaling issued by the International Whaling Commission (IWC) and enacted in 1986. It does not matter that the campaign will soon be 50 years old; the slogans and imagery remain fresh and familiar.

This linkage between local protests and international action may seem difficult to replicate. Many hundreds of problems have sought action at the international level without achieving any meaningful victories. Nevertheless, international meetings dictate the levels of protection (or lack thereof) that people worldwide enjoy. International environmental treaties include provisions that seek to limit the ozone hole's size, discourage international trade in endangered species, and prevent risks to public health and the environment stemming from the transportation of hazardous waste.

Perhaps not as obvious, international environmental issues also include social concerns like environmental justice, the distribution of environmental burdens and benefits around the globe, access to education and jobs, and gender equality. There is a recognition that poverty, including the absence of economic activity, also contributes to environmental damage in both the natural and built environment. The so-called North–South gap encapsulates the inequalities between the developed Northern countries and Southern developing countries. This gap also shapes international environmental affairs as both hard and soft law must reconcile the differences in norms, causes of devastation, preferred policies, and treaty contents between these developed and developing countries.

Whether the issue of concerns includes traditional focuses of environmental conservation and elimination of pollution or more recent topics like climate change and sustainability, the need for an international approach to environmental problems appears, at first glance, to be apparent. Environmental pollution travels across national boundaries in an ever-greater reach, impacting populations without the opportunity to determine their environmental health. Air pollution with its origins in China may well contaminate the air within the United States. Similarly, air pollution that originates on the East Coast of the United States settles in part on Western Europe. Today, long-range transport of chemicals makes their way around the globe, including to the Arctic, where these chemicals may bioaccumulate [16].

The movement of pollution across national boundaries is not limited to the long-range transport of air pollution, as water pollution also exemplifies the transboundary nature of environmental damage. The Danube River, originating in Germany, passes through ten countries in Central and Eastern Europe before entering the Black Sea in Romania. The Danube River suffers from excessive nutrient loading, organic material, and hazardous chemicals, including antibiotics and microplastics. These materials enter the water from many farms and factories in one of the nineteen countries that drain into the Danube River. In response to environmental damages, these countries formed a new international organization, the International Commission for the Protection of the Danube River. This organization seeks to foster better communications, and more importantly cooperation, to limit damage to those who depend on the river system, whether industrial manufacturing processes, agriculture, livestock, humans, or wildlife.

Thus, environmental damage occurs worldwide, even if there is no immediate source of chemical contamination. Living in a pristine environment is no longer possible. All people face environmental risks from the current organization of society. This environmental risk is itself unequally distributed across the globe. Differences in actual or perceived ecological damage and differences in risk tolerances make it more challenging to implement actions to end environmental damage since the distribution patterns of pollution require worldwide cooperation to succeed.

States may, and frequently do, exercise their option to create, alter, or abolish international organizations to lessen and reverse environmental pollution. In the last fifty years, states created new international organizations, either as standalone entities or as part of the United Nations (UN) system, to enhance cooperation while promoting global environmental protection. Additionally, states also added environmental protection to the agenda of a growing international arena.

Despite its status as one of the minor issues compared to military affairs or trade negotiations, environmental policy enjoys significant status as a subfield within international relations. The field continues to yield promising results for understanding the hows and whys of global politics. Eschewing traditional theories of power politics and economic dominance of states, the academic subfield of environmental affairs instead expands into the realm of the non-state actor, knowledge, influence, and normative values and beliefs. Despite the noticeable absence of powerful militaries and significant wealth, pro-environmental groups maximized their ability to shape and change global diplomacy on behalf of the environment.

1.1   The International State System and the UN

While it seems natural and inevitable today, the state system did not always exist. Historians date the international system to the Treaty of Westphalia in 1648 as containing the origins of state sovereignty, the state's right to choose its fate without outside influence. This treaty ended a conflict between the Holy Roman Empire and the early European states that saw secularism replace religious authority as the basis for organizing society. The great European empires of Austria, Russia, Prussia, England, France, and the areas that would become The Netherlands and Belgium, established the concept of state sovereignty, meaning that the state controlled its territories, free from the interference of outside influences.

The state system proved to be relatively stable but frequently violent. States pursued their interests, including attacking other states. The states also developed centralized control over their territories so that over time, the state became the primary voice of its citizens in representing their needs, wants, wishes, and demands. Thus, states gained legitimacy, that is, the consent of the governed, in representing their citizens.

The European empires remained at the heart of the international state system through the next two centuries, establishing powerful empires that controlled the remainder of the world. Indeed, British citizens proudly proclaimed theirs was the empire on which the sun never set. While colonies located in the new world successfully revolted and established their own independent state, areas closer to Europe, such as Africa, remained trapped in the colonial system.

The expansion of the colonial empire system increased the political and military tension around the globe and in Europe as the empires competed for territory and resources throughout the world. The so-called Concert of Europe attempted to manage this system by engaging in bilateral and multilateral treaty negotiations to reassure nervous states of their place in the world. However, this system broke down upon the assassination of Archduke Franz Ferdinand in 1914 when a series of hidden alliances negotiated by the great empires expanded a regional conflict into a world war.

In the aftermath of World War I, the Americans, at the urging of their President, Woodrow Wilson, sought to implement a revolution in how states managed their affairs. States intended the League of Nations to serve as a public forum for states to address their grievances. However, the United States failed to join as a member state. Ultimately, the League of Nations suffered from poor institutional design, as it required unanimous consent from its members. The Axis powers, Germany, Italy, and Japan eventually withdrew their League membership.

At the end of World War II, the state system formalized relations between other states by creating the UN. The two great wars fought in the first half of the twentieth century firmly entrenched the idea that some type of permanent neutral structure could reduce further violent conflict. After the Allied military victories that heralded the end of World War II, these powers negotiated a new system to maintain the international state system.

Led by the United States, the Allied powers created the UN by negotiating the UN Charter. Many international treaties, including the UN Charter, open for signature, followed by a ratification process. The UN Charter opened for signature on June 26, 1945, at the San Francisco Conference. Ratification occurred on October 24, 1945, with the approvals of the five permanent members of the UN Security Council (the Soviet Union, France, China, the United Kingdom, and the United States) and a majority of the remaining countries in existence at the time. These states charged the UN with preventing a future world war by maintaining communications and resolving conflicts between the nations. This operational goal also means that the UN promotes cooperation and seeks to minimize inequalities that could lead to war in the future, including access to resources such as food, water, shelter, and medicine.

The UN Charter provided an internal structure to the organization, including the General Assembly (UN GA), the Security Council, the Economic and Social Council (ECOSOC), the Secretariat, the International Court of Justice (ICJ), and the Trusteeship Council. The UN GA functions as the primary policy-making body for the UN. Each member state may introduce resolutions for consideration, with adoption by a simple majority vote. The UN GA also holds the ability to convene international negotiating committees and to give direction to the specialized agencies and other bodies within the UN system.

The Security Council consists of fifteen member states within the UN. This body works to maintain peace by investigating situations that threaten to escalate into violence. The Security Council members may impose sanctions on states engaging in conflict. Five of these countries, held and continue to hold permanent membership based on their alliance at the end of World War II, while the remaining ten member states serve two-year terms upon election. Additionally, any one of the five permanent members may also veto measures proposed by the Security Council, even if the remaining fourteen members support the measure.

ECOSOC works to coordinate the various components of the UN system, especially with regard to economic and social issues. Today, sustainable development, a critical paradigm within international environmental politics, serves as the central concern for this council. ECOSOC consists of 54 member countries elected to the council on three-year overlapping terms. Seats are assigned to the official regional groups to maintain political balance. In addition to coordinating the actions of the UN system, ECOSOC also accredits observers to the UN, including NGOs. As of 2021, 4045 NGOs maintain consultative status [25].

The Secretariat serves as the executive office for the UN system. Headed by the Secretary-General, this office provides day-to-day oversight for the thousands of employees and peacekeepers working at this international organization. The Secretary-General serves a five-year term upon election by a majority of the countries with membership. Of the nine men occupying this office, four have been from developed countries, five from the developing South. Mr. António Guterres of Portugal currently occupies the office. The Secretariat contains a variety of departments and offices, organized by function. New York City hosts the UN headquarters, with other important offices in Geneva, Switzerland, Nairobi, Kenya, and Vienna, Austria.

The ICJ, located in The Hague, Netherlands, functions as a legal arbitrator between states. The justices of this court settle disputes based upon the principles enshrined in international law. Fifteen judges sit as a panel to hear disputes, and no more than one judge may originate from the same country. Any state that is a member of the ICJ may bring a dispute before the court. The ICJ reviews cases under both a broad compulsory function and when specified in an individual treaty. Some states do not participate in the ICJ. Thus, the membership in the ICJ differs from that of the UN GA. The United States, in particular, does not accept the compulsory authority of this court.

The Trusteeship Council initially held a significant role within the UN system. This council oversaw the creation of independent countries after the break-up of the colonial system at the end of World War II. The great empires released their colonies or, in the case of the defeated Axis powers, surrendered their colonial territories into the International Trusteeship System with the goal of creating self-governing, independent countries. After the last territory exited the International Trusteeship system in 1994, the Trusteeship Council changed its meeting frequency as demand for its services dropped dramatically.

In addition to the six main organs within the UN system, various additional agencies joined the UN system. These specialized agencies have a unique and varied history. For example, the International Labor Organization (ILO) works to create a uniform and just labor standards, and the International Telecommunications Union coordinates communication systems such as the telegraph, radio, and telephones, both predated the UN system. States also created new specialized agencies like the Food and Agricultural Organization (FAO) that oversees work relating to ending hunger and improving agricultural productivity.

With the lessons learned from the failed League of Nations, member states collectively engaged in the UN, with the strong leadership of the United States. The need to manage the decolonization process and the rise of the Cold War tensions between the Soviet Union and the United States gave the UN more relevance than its predecessor organization. Over time, the UN's mission transitioned from a singular focus on peace and security to focusing its efforts on humanitarian issues, including economic and social development.

While states dominate the UN system, observers participate in many negotiating sessions. Informally, non-state actors may meet with negotiators in an attempt to influence the outcome of the meeting. Additionally, non-state actors tend to form coalitions with like-minded groups at the international level. Thus, environmental NGOs merge their messages in the hopes of attracting greater attention and political influence. These transnational advocacy networks [21] form across national boundaries. With communication aided by a multitude of social media tools, these transnational advocacy networks shape public opinion globally, increasing the environmental movement's ability to apply political pressure to governments and corporations.

In addition, NGOs express their thoughts in meetings. They publicize the conference by engaging with the global media, and they inform their members concerning the ongoing status of the negotiations. Today, NGOs not only participate inside the meeting hall, they also create spectacle outside the conference by organizing protests and marches in hopes of creating publicity in favor of environmental protection.

1.2 Organizing Themes

As we near the fiftieth anniversary of the original environmental conference, the time is ripe to review the origins and impacts of the role of environmental affairs within the international system and its attempts to advance environmental protection on behalf of states and their citizens. Environmental policy falls under the heading of the so-called wicked problems. These problems are complex, intractable, and intensifying in their impacts on society [6, 9, 18, 19, 24]. Wicked problems by nature lead to complex international diplomatic interactions between various actors. Thus, this book analyzes environmental politics along three fundamental conceptualizations; complexity, change, and continuation. These three conceptualizations assist in ordering information about international environmental diplomacy. While Keohane and Nye [22] also acknowledge these characteristics as hallmarks underlying the international system, this work also incorporates the physical world.

Complexity occurs for a variety of reasons inherent to the environment. Transboundary movement of pollution repeatedly occurs as environmental media such as air, water, and soil are interconnected in practice within an ecosystem. Complexity arises for a variety of reasons. Air, water, and soil are connected within the global ecosystem. Pollution does not stay neatly confined within environmental boundaries or national jurisdictions. Sulfur oxides and nitrogen oxides enter the atmosphere as air pollution but may damage water and soil as these chemicals react to form acid rain precipitation before returning to earth as part of the water cycle. More insidiously, attempts to clean one media, for example, air, may result in damage to another media, soil, through an increase in hazardous waste that may be placed in a hazardous waste landfill. Scrubbers, a common control technology used to eliminate air pollution from burning fossil fuels such as coal, create the need to dispose of the waste chemicals from utilizing the scrubber. This creates greater amounts of hazardous wastes that must be properly disposed of [23]. This complexity compounds itself as the pollution moves across ecosystems, regions, and national borders.

An essential concept within environmental affairs, leakage occurs when an environmentally destructive action is relocated. Industrial sites that create pollution have relocated to avoid placing scrubbers as these pieces of equipment significantly increase the costs to run a facility. The original location gains a cleaner environment but suffers social costs as the jobs at the facility end. Leakage not only connects impacts at one place to another, but also it connects seemingly unrelated issues.

One of the more intractable problems within international environmental diplomacy involves the relationship between the environment and the economy. Within the United States and other developed nations, citizens often view industrialization as the root cause of ecological devastation [12, 14]. However, other parts of the world view underdevelopment as the primary environmental problem. In other words, people and societies without access to adequate shelter, sanitary systems, clean drinking water, or adequate supply of fuel for cooking and heating also become exposed to environmental health problems, increased risk exposures, and experience environmental damage [13].

International environmental diplomacy recognizes both causes, industrialization and lack of access to technology and infrastructure needed for an adequate quality of life as part of the international environmental agenda. Consequently, the proposed solutions to ecological disorganization may also be different. In the case of the industrialized North, technological preferences for the so-called end-of-pipe solutions that mitigate environmental pollution dominate. However, in the developing South, redistribution of industrialization, including technology transfer, knowledge transfer, and increases in development aid from North to South, also occur as part of the environmental agenda.

One of the fundamental conflicts within this realm includes the role of energy. Industrialization historically depended on fossil fuels, primarily coal, but also oil and natural gas. The combustion products of these fuels lead directly to the increased concentration of greenhouse gases in the atmosphere. As these gasses accumulate, the entire earth's systems also change, leading to a rash of unwanted impacts, including increased natural disasters and sea level rise. Since the fossil fuels that power much of the international economy create carbon dioxide, limiting the damage to the planet relies on finding new affordable sources of energy. Ironically, developed countries that have utilized fossil fuels to develop are now converting their energy systems to renewable energy as part of their environmental plans. In contrast, developing countries cannot afford to switch to renewable sources.

There are very few universally accepted environmental treaties, even though virtually every government participates in environmental negotiations. International diplomacy is a slow, drawn-out process that favors the inputs from national governments. The 193 member states of the UN write the rules for interactions at the global level. These rules do not have to consider environmental protectionism as states choose which elements of their domestic societies they emphasize while attending UN meetings.

Reconciling these differing stances requires time, fortitude, and patience. Unfortunately, environmental harms continue to accrue while these diplomatic processes implement conflict resolution. In the meantime, economic interests benefit from protracted negotiations that postpone changes in the way society functions. Further, these competing interests create a piecemeal system of overlapping international treaties that allow for ongoing practices to continue. Increasing the number of accords and lengthy negotiating time frames potentially complicate domestic implementation and enforcement activities. States and corporate actors may have difficulty in implementing and complying with the various requirements within the treaties due to the decentralized nature of the entire environmental system.

The complexity of the international environmental agenda adds to the difficulty in understanding the overwhelming number of environmental treaties. The global environmental agenda sends conflicting signals about the underlying social values that states, corporations, and individuals should follow. The polarization between the economic growth mantra and efforts to protect the environment exemplifies this dichotomy.

In addition to the other complicating factors, scientific cause and effect may also create complexity as science is a dynamic discovery process. Despite the scientific advancements in the post-World War II era, environmental scientists continue to generate new insights into how the environment functions. These insights involve new systems thinking that focus on the interconnectedness of many individual parts across both time and space.

Given the severe and significant nature of environmental problems, the continuation of "business as usual" may not be in the best interest of a global society. If wicked problems demand unique solutions, focusing on change mechanisms must be a key component of international environmental studies. Planetary impacts require planetary changes. Thus, all levels of governance, from the local to the global, must engage simultaneously. Doing so will be neither simple nor straightforward.

Diplomatic interactions do not remain constant over time; change and variation become increasingly important, especially as environmental affairs age. How one accomplishes change should be considered one of the central tenets within environmental politics. This arena begins with the fundamental tenet that the consequences of environmental damage are sufficiently severe that environmental damage must end. This cessation of activities does not restore the environment to its original condition. It only stops future damage from occurring. A "deeper green" position might assume instead that environmentally damaging activities must end and that the problem definition should also include environmental restoration [17].

Political science scholars study the ability of an actor to impose their will on another by examining key concepts such as power and influence [3, 5, 10, 11]. While scholars have failed to reach a consensus regarding a precise conceptualization for power, this concept remains at the heart of the discipline. The relationship between power and influence assists in analyzing behavior change. Many political scientists separate these two concepts, with power emphasizing capabilities [10]. Power constitutes all of the political resources held by a state. Influence then constitutes a change of behavior due solely to the actions of the actor agitating for change [5].

In addition to conceptualizing change behaviors, scholarly models of international relations theory incorporate change processes differently. International relations theory utilizes competing theories to analyze global events in both the present and the past. This manuscript focuses on four primary international relations theories along with other theories that seek to explain international environmental diplomacy. Each of these theories highlights different elements within the international state system and illustrates important strengths and weaknesses of international environmental diplomacy.

Despite the attempts to reform the international state system to incorporate environmental concerns, the diplomatic system nevertheless contains a vital element of continuity. This continuity occurs due to the dominance of economic norms and the stability of the international state system that prizes state sovereignty.

In practice, continuity between issue areas within environmental diplomacy occurs with regularity. Diplomats tend to represent a country in more than one negotiation. For example, Ambassador Tommy Koh of Singapore served as the President of the Third UN Conference on the Law of the Sea (UNCLOS III) in 1982. At the UN Conference on Environment and Development (UNCED) in 1992, Ambassador Koh chaired the main committee at this meeting. Perhaps more importantly, soft law principles, the guiding ideas embedded within the treaties, frequently carry over from one treaty negotiation to the next. One of the essential environmental soft law principles is common but differentiated responsibilities. This principle affirms that all countries are bound together as residents of the same global environment while also acknowledging that different countries have different resources and abilities to make changes to protect the environment. In other words, the United Kingdom, as a formerly great empire and wealthy nation, has more responsibility to protect the environment than a country such as the Maldives. The Maldives, an island nation in the Indian Ocean, did not contribute to environmental damage as the residents of the United Kingdom. Equally relevant, the Maldives does not have the same financial resources to protect the environment like the United Kingdom.

While seemingly contradictory that this book focuses on both change and continuity simultaneously, the complexity of the international system may well impact the continuity of the system itself as large systems with a high number of independent actors remain difficult to change. This difficulty increases when the degree of change increases. Thus, practitioners seeking to change the entire international state system from a profit-driven system that relies on state sovereignty to one that focuses on sustainable development with an emphasis on equity between peoples in both time and space may not be able to change the system simultaneously.

1.3 Purpose of the Book

It is not the purpose of this book to comprehensively document every twist and turn within international environmental diplomatic history for the simple reason that no single book would be long enough to complete this task. Nor is it the purpose of this book to solve the great debates of political science. It is enough, for now, to layout, in a cohesive, systemic way, the chronology of events and the political science theories that these events inspired.

This manuscript instead examines, in chronological order, headliner international conferences beginning with the 1972 UN Conference on the Human Environment (UNCHE). Thus, this book utilizes an explicitly interdisciplinary approach informed by international relations, international law, environmental studies, and environmental science. This book is certainly not the first text with this noble goal in mind. Global environmental politics textbooks frequently seek to explain the importance of international environmental issues by assuming that students have a significant political science background in the international state system. In doing so, texts such as [2, 8] briefly mention significant events before focusing on contemporary controversies, all the while explaining political science concepts. In contrast, this text seeks to simplify this approach by providing a broad overview of events before presenting the scholarly work that defines the international environmental affairs subfield. Thus, undergraduate students new to international environmental issues may find this book helps fill in the blanks unintentionally created by focusing exclusively on the international relations lens. In other words, this book seeks to be explicitly interdisciplinary in outlook, rather than adopting a theoretical approach such as liberalism while studying environmental politics.

The book also expands the emphasis beyond personal advocacy as some portion of this book's audience may find a career in an area that requires a working knowledge of international laws and diplomacy that necessitates a different type of sophistication beyond understanding pathways for advocacy. The global environmental realm generates legal obligations and soft law principles that major corporations and small businesses benefit from supporting voluntarily. The growing number of treaties and overlapping legal regimes raise the stakes of mandatory compliance, monitoring, recordkeeping, and reporting, and will continue to provide high quality jobs for workers around the globe.

Each chapter begins with an overview of the issue area by examining the environmental problem, potential solutions, and the process by which diplomats negotiated agreements. Each episode starts by reviewing the negotiating history of the issue area, along with the treaty. The chapter then turns to a political science-oriented review of the historical context. The processes by which countries cooperate change over time. Thus, each issue area prompts new analyses that individually and collectively give insight into our current international governance systems.

The next section of the book begins our examination of the rise of international environmental issues by looking at the origins of the international environmental framework. We start by looking at the 1972 UNCHE in Chap. 2, while Chap. 3 reviews the establishment of the United Nations Environment Program (UNEP). Chapter 4 reviews common pool resources such as whaling, the high seas, and oceans. Chapter 5 looks at the Montreal Protocol on Substances that Deplete the Ozone Layer. Chapter 6 concludes this section by reviewing the Basel Convention on the Control of Transboundary Movements of Hazardous Wastes and their Disposal (Basel Convention).

The maturation of international environmental diplomacy forms the second section of the book by focusing on the Earth Summit and its Aftermath in Chap. 7. Chapter 8 looks at the UN Framework Convention on Climate Change, while Chap. 9 reviews the Convention on Biological Diversity (CBD) and its protocols. Chapter 10 completes the review of trade in chemicals and hazardous waste by reviewing the Rotterdam Convention on the Prior Informed Consent Procedure for Certain Hazardous Chemicals and Pesticides in International Trade (Rotterdam Convention) and the Stockholm Convention on Persistent Organic Pollutants (Stockholm POPs).

The book's third section examines sustainability in the twenty-first century, beginning with the Millennium Development Goals (MDGs) in Chap. 11. Chapter 12 looks at the World Summit on Sustainable Development (WSSD) that failed to add new principles or provide a broad vision for sustainability but promoted the role of non-state actors. Chapter 13 reexamines climate change in the absence of new carbon reduction targets. Chapter 14 reviews the 2030 Agenda for Sustainable Development (2030 ASD) and the Sustainable Development Goals (SDGs). Concluding thoughts close out this manuscript in Chap. 15.

As the history of global environmental governance unfolds in this book, I hope to promote great hope and optimism for the future. More people in more places today work on solving these wicked problems than at any time in the past. The so-called "hard" sciences reveal new insights and information about how the earth functions, allowing today's citizens to create new solutions to both old and new problems. Innovations in technologies and communications facilitate cooperation among activists, scientists, and diplomats in ways which were inconceivable in the past. Similarly, once the realm of elder statesman, international diplomacy transformed into a diverse, vibrant community that better reflects the difficulties of ordinary peoples around the world. Incorporating these voices can only lead to a more dynamic and productive conversation capable of protecting the planet in the future.

References

1. Alsharif M, Andone D (2019, October 10) Climate change protestors with a boat blocked traffic today in New York's Times Square. CNN https://www.cnn.com/2019/10/10/us/extinction-reb ellion-times-square-trnd/index.html. Accessed 19 Nov 2021
2. Axelrod R, VanDeveer SD (eds) (2020) The global environment: institutions, law and policy. CQ Press, Thousand Oaks
3. Arts B (1998) The political influence of global NGOs: case studies on the climate and biodiversity conventions. International Books, Utrecht
4. Berlin DPA (2019, October 9) Dritter Protestlag: extinction rebellion besetzt Marchallbrücke. [Press Release] https://www.berlin.de/aktuelles/berlin/5933074-958092-extinction-rebellion-protest-marschallbr.html. Accessed Nov 19 2021
5. Betsill MM, Corell E (2008) NGO diplomacy: the influence of non-governmental organizations in international environmental negotiations. MIT Press, Cambridge
6. Brinkmann R (2020) Environmental sustainability in a time of change. Palgrave, Basingstoke
7. Busby M (2019, October 3) Extinction rebellion protestors spray fake blood on to Treasury. US Edition, The Guardian. Retrieved from https://www.theguardian.com/environment/2019/oct/03/extinction-rebellion-protesters-spray-fake-blood-treasury-london. Accessed Nov 19 2021

8. Chasek PS et al (2018) Global environmental politics. Routledge, New York
9. Churchman CW (1967) Free for all. Manage Sci 14:B141–B142
10. Cox R, Jacobson H (1973) The anatomy of influence: decision making in international organization. Yale University Press, New Haven
11. Dahl RA (1957) The concept of power. Behav Sci 2(3):201–215
12. Dalton RJ (1993) The environmental movement in Western Europe. In: Kamienicki, S Environmental politics in the international arena: movements, parties, organizations, and policy, pp 41–68
13. Defries RS (1983) The role of environment in the development process. Intl Bus Law 11(2):52–54
14. Dunlap RE, Scarce R (1991) Poll trends: environmental problems and protection. Public Opin Q 55(4):651–672
15. Extinction Rebellion (2019, November 6) High Court rules London Protest Ban Unlawful. Retrieved from: http://bbc.com. Accessed 19 Nov 2021
16. Gibson JC (2020) Emerging persistent chemicals in human biomonitoring for populations in the Arctic: a Canadian perspective. Sci Total Environ 708:134538
17. Hall M (2005) Earth repair: a transatlantic history of environmental restoration. University of Virginia Press, Charleston
18. Head BW (2008) Wicked problems in public policy. Public Policy 3(2):101
19. Head BW, Alford J (2015) Wicked problems: implications for public policy and management. Admin Soc 47(6):711–739
20. Kalland A (1993) Management by totemization: whale symbolism and the anti-whaling campaign. Arctic 46(2):124–133
21. Keck ME, Sikkink K (1998) Activists beyond borders: advocacy networks in international politics. Cornell University Press, Ithaca
22. Keohane RO, Nye JS (1977) Power and interdependence: world politics in transition. Little Brown and Company, Boston
23. Moretti AL, Jones CS, Asia PG (2012) Advanced emissions control technologies for coal-fired power plants. Power-Gen Asia, Bangkok, Thailand
24. Rittel HWJ, Webber MM (1973) Dilemmas in a general theory of planning. Policy Sci 2:155–169
25. UN DESA (2021) How to apply for consultative status with ECOSOC. https://www.un.org/development/desa/dspd/civil-society/ecosoc-status.html. Accessed 20 Nov 2021

Chapter 2
International Environmental Diplomacy Begins

Abstract The United Nations Conference on the Human Environment propelled international environmental affairs into the global spotlight. States meeting at this conference created the Stockholm Action Plan to create new international infrastructure to support states' efforts to protect and preserve the natural and built environment. This process, however, was not straightforward as the North–South gap emerged over differences in international priorities. Northern states sought collaboration to limit environmental damage without causing economic losses, while Southern states sought to increase industrialization largely viewed as the root cause of pollution in order to improve their economic standing and overall quality of life. In order to reach compromise, proponents of the Stockholm Conference created the concept of eco-development, the intellectual predecessor of sustainable development to acknowledge that Southern developmental concerns should be addressed as part of the environmental agenda.

Keywords United Nations Conference on the Human Environment · North–South gap · Stockholm Action Plan · Eco-development · Non-governmental organizations · Stockholm Conference

This chapter begins the inquiry into the overall trajectory of international environmental politics by looking at the UNCHE negotiating process and the events that triggered this conference. Ecological problems have existed throughout the entirety of human history. London, in particular, suffered through a series of air pollution episodes dating back to the medieval era due to the burning of coal [5]. London's worst air pollution episode, the Great Smog of London, occurred in 1952 when high coal usage happened on a cold, windless day. Air pollution from particulate matter was trapped over the city, and acid rain fell. The resulting air pollution lasted five days from December 5–9, 1952, causing numerous people to suffer from respiratory illnesses. While the number of people impacted will forever remain unknown, the estimated death toll easily numbered in the thousands. Health officials at the time estimated that 4000 people died immediately, with later studies suggesting the death toll might have been closer to 10,000–12,000, including deaths from indirect health impacts [2].

The incidence, frequency, and severity of air pollution closely follow the rise of fossil fuels such as coal, fuel oil, and to a lesser extent, natural gas. This story of local smog episodes continues in modern industrialized society. It should not be surprising that concern over environmental damage, including acid rain, drove countries to discuss ways of cooperating to minimize ecological decline after scientific research concluded that air pollution caused local problems and crossed national boundaries. Nations that received pollution began the arduous process of adding these issues and other transboundary environmental issues to the international agenda. This international action occurs in addition to national attempts to control pollution.

Thus, when Sweden discovered that acid rain with both local and international sources was causing environmental destruction, its diplomats called for a conference to discuss environmental degradation and global solutions in 1968 by submitting a resolution to the UN GA. Scholars of international environmental diplomacy typically recognize this resolution as the beginning of the ascendency of this issue area on the global agenda. In the first twenty years of international environmental diplomacy, international activity began with the need to raise awareness of the commonalities of local environmental destruction by hosting conferences and finalizing treaties. To achieve these goals, the UN engaged national governments. Conference organizers also discovered the value of adding interested observers, formally called NGOs, to the international treaty making process. This terminology serves as a catch-all category that lumps pro-environmental groups with non-profit organizations representing businesses and industries. This expansion from the state to the non-state raised the awareness of the global public. It also created an opportunity to interact with and openly support both the international diplomatic processes that create international treaties and the implementation of the finalized treaty text at all levels of government.

Consequently, this chapter reviews the first mega-conference in international environmental affairs, the UNCHE, held in Stockholm, Sweden, on July 5–16, 1972. The so-called Stockholm Conference occurred in a time of substantial conflict. States grappled with the East–West conflict of the Cold War and the beginnings of the North–South gap. While the struggle between the United States and the Soviet Union formed the major military and economic conflict in the 1970s, the North–South gap denotes a diplomatic impasse over access to wealth, technology, and the use of natural resources. Current practice splits countries into two groups based upon their degree of industrialization, including economic wealth, technological advancement, and level of education, with the United States leading the North Atlantic Treaty Organization members and the former Warsaw Pact members in the North. All the other countries identified with the developing South.

The first section reviews the beginnings of international environmental politics originating from the lackluster state of the global environment in 1968. The second section examines Sweden's request to the ECOSOC to host a conference and the preparatory committee (PrepCom) process focusing on the significant schisms between countries from 1969 and the opening day of the meeting in 1972. The third section focuses solely on the UNCHE formal negotiating session to see how international diplomacy handled these tensions. This section also includes a brief review of the significant outcomes of this mega-conference. The fourth section ends with a review of the contemporary scholarly literature assessing the event and its immediate aftermath.

2.1 State of the Global Environment

All cultures, nations, and states have environmental struggles, and each one of these institutions must decide its relationship with the human environment as the two are intrinsically linked. When the decisions of either the individual, the culture, the nation, or the state negatively impact others, then international diplomacy becomes more viable as traditional principles of international law such as "do no harm" and "good neighborliness" dictate the responsibility of the polluting state to keep the receiver of the pollution "whole." However, these two legal principles do not require the polluting country to eliminate all pollution. Legal jurisprudence requires only that the dispute is settled to the satisfaction of both parties, including the possibility for a transfer of funds or technology in compensation for damages done.

A relationship between environmental damage and the types of industrial activities does exist as chemical processes tend to cause similar damages based on the quantity of production output and manufacturing processes used. Thomas Malthus expressed the earliest version of this work in his 1798 work, *An Essay on the Principle of Population*. One of our enduring environmental philosophies, this theory postulates that exponential population growth will overwhelm linear increase in food production, causing a Malthusian catastrophe [20]. As such, humans would always be subject to misery. One application of this theory suggests that an expanding population will cause consumption to increase exponentially with a corresponding explosion in ecological destruction.

Early environmental movement leaders Paul Ehrlich, Barry Commoner, and John Holdren developed a more recent expression of the relationship between population and consumption in the early 1970s, the so-called "IPAT" formulation. The IPAT equation states that impact, or carrying capacity, is a function of population, affluence, and technology.[1] While not mathematically correct, this simple formulation illustrates the complexities that link industrialization, wealth generation, and the quality of natural and built environments together. While perhaps more optimistic than pessimistic Malthus, the IPAT formulation intends to suggest that there are limits to growth. Thus, for Malthus and his supporters, population control becomes necessary for humankind's present and future well-being to ensure that everyone has a reasonable quality of life with access to affluence and technology.

It is not surprising that major concerns with industrialization in this time frame focused on the impact of pesticides, oil usage, and conservation of natural resources. A variety of media outlets, including book publishers and print journalists, brought these concerns to the public's attention. The New Yorker magazine initially published

---

[1] For a comprehensive treatment on the intellectual development of the $I = PAT$ formulation, see [7].

Silent Spring in June 1962 as a three-part series before Houghton Mifflin consolidated the articles into a book in September of that same year.

Rachel Carson's *Silent Spring* poignantly discussed the impact of dichlorodiphenyltrichloroethane (DDT) and other pesticides on the environment. Carson extensively documented the many harms DDT caused insects, birds, other wildlife, and people [6]. The book concluded that continued pesticide use would cause widespread damage to people and the environment. The public outrage over these damaging impacts eventually led to the banning of DDT in many Northern countries. It also contributed to the creation of the Environmental Protection Agency (EPA), the United States government agency charged with protecting the environment. Created by executive order on July 9, 1970, and beginning operations on December 2, 1970, this agency became one of the first attempts to adapt governmental structures to oversee the response to environmental pollution problems. The EPA would, in turn, become the blueprint for many other countries organizing to participate at the Stockholm Conference.

Television stations likewise covered environmental affairs as part of the nightly news. Environmental icons entered pop culture representing both the highs and the lows of technological advancement as evidenced by events such as the 1967 Torrey Canyon oil spill and the Spaceship Earth picture, Earthrise, in 1968. With the recognition that various industrial and agricultural activities could impact ecological systems, people worldwide realized that environmental damage was not limited to their immediate surroundings but was a shared problem that required cooperative solutions. Torrey Canyon, an oil supertanker, wrecked off the coast of Cornwall in the United Kingdom in 1967. The resultant oil slick reached the shores of both England and France. The images of the British Navy repeatedly torpedoing the ship in an attempt to burn the oil off at sea, rather than allowing the oil slick to reach land, emphasized the interconnectedness of the natural environment as well as humans' poor control over anthropogenic environmental damages. Torrey Canyon, in particular, caused an outpouring of public support for international environmental protection. Equally importantly, the Torrey Canyon oil spill spurred citizens to question who was responsible for the clean-up in the global commons, the four areas (oceans, atmosphere, Antarctica, and outer space) where no country has jurisdictional control.

While Torrey Canyon illustrated the fragility of our planet, Spaceship Earth illuminated its innate beauty and connectedness. This picture of Earth suspended in space from the Apollo missions conveyed the idea of the limits of nature as well as a need to protect the biosphere. The images from the Apollo program deeply impacted environmental consciousness and entered pop culture with multiple movies, books, and speeches referring to the concept of a Spaceship Earth. This pop culture symbol of the late twentieth century provided a theme for new awards, hundreds of books, and speeches. Eventually, it became the name of a ride at EPCOT, part of the Walt Disney World vacation complex outside Orlando, Florida. In doing so, Spaceship Earth came to represent the conceptualization of Earth as possessing finite resources that must be conserved for the betterment of humanity. The Spaceship Earth imagery captured the general public's imagination, leading to greater interest in environmental affairs.

It moved beyond the scientists and researchers whose place in education gave them their insight into the cause and effects of science, technology, and the environment.

In addition to these three significant events, many other environmental disasters left visible environmental damage in every industrial country. Whether from oil spills or pesticide use, citizens joined together to combat the industrial assault on the environment, regardless of whether this assault was purposeful or accidental. In the democratic West, free peoples demanded action to articulate and implement new laws on environmental protection at home. They also supported international discussions that could lead to cooperative efforts to eliminate pollution from sources abroad. However, the situation in the East did not focus on either domestic or international cooperation. Many of the communist governments in the East refused to admit that communist economies, no less than capitalist economies, caused pollution.

Sadly, these communist regimes would not tolerate any local actions contradicting the official government's position. This resulted in limits to community activities, including environmental organizations that prodded governments into taking environmentally protective actions. Further, governmental orders to undertake actions that damaged the environment would be obeyed without question.

Environmental concerns vary based upon global position. The global South represents the states dominated economically, politically, and culturally by Europe and North America. This colonization started in the fifteenth century and continued officially until the decolonization process that began at the end of World War II in 1945. The Southern coalition, alternatively known as both the non-aligned movement or the Group of 77 (G-77), was initially formed to deal with decolonization, apartheid, poverty, and future economic development prospects. With its semi-militant mantra and demonstrating a willingness to reshape the UN via control of the UN GA with its one nation, one vote rule, the coalition sought to avoid entanglement in the Cold War while simultaneously promoting their economic development. The G-77 should not, however, be thought of as a homogenous bloc as it routinely subdivides along regional, cultural, and economic lines. Today, 134 states are members of the G-77.

The G-77, then, remained unconvinced that environmental affairs ought to be an item on the international agenda. Using the argument that industrialization causes increases in environmental pollution, this argument surely did not cover the South as they had little industrial development. Southern states focused on the lack of clean water and the need for improved housing and basic sanitation. These states also sought to enhance their quality of living, either through receiving higher payments for raw materials or through access to technology to produce finished goods in the South.

2.2 The Conversation Begins

Diplomats waded into this fractured ideological debate on whether to convene a convention to discuss the human environment when Sweden forwarded a letter to ECOSOC on May 20, 1968, stating that only international cooperation could solve ongoing manmade changes to the environment [9]. In no small part, Sweden did this to focus attention on the amount of acidic precipitation (acid rain) that fell on their country from European sources outside of its jurisdiction. Other countries agreed that pollution did not respect borders, and widespread support for an international conference soon followed. Thus, member states of the UN began their efforts in planning and, after accepting an invitation from Sweden, convening the UNCHE in Stockholm from June 5–16, 1972. As stated previously, the scope of the meeting incorporated a broad framework that included air, water, and soil pollution, biocides, and the conditions of man that could be impacted by the environment, including working conditions and quality of life concerns. Through its diplomat at the time, Sverker Åström, Sweden suggested that potential treaty outcomes could include a convention on oceans, freshwater resources, or prevention of air pollution [28].

At a UN conference, states hold a preeminent place within the international system. The sovereign state controls, at least in theory, if not in practice, a fixed amount of territory and all activity within that territory through domestic laws and regulations. The state also holds exclusive use of force through either its police or its military. All other types of violence are seen as illegitimate, whether a crime, domestic terrorism, or external invasion. States chose when and where to cooperate on a specific issue area. Thus, UNCHE also marks an expansion of individual state's agreement to cooperate on international environmental affairs.

From the onset, the UNCHE endeavor needed to overcome two significant hurdles typical of politics during the Cold War era. The first involved the extent to which conference proponents could secure funding. The second involved Cold War politics, in this case, East Germany's ability to attend the meeting. East Germany, more formally known as the German Democratic Republic, did not have membership in the UN, nor was it a member in a specialized agency. Interestingly, scholars provided slightly different lists of countries that boycotted the Convention due to the status of East Germany; the consensus that emerges from these resources concludes that Romania may well have been the only member of the Soviet bloc to attend the Stockholm meeting [12, 19, 25].

Secondary to the question of East Germany's status, another non-trivial claim about the absence of a relationship between the communist market system and environmental damage is also worth mentioning, if only to elucidate and debunk the myth that communist countries have minimal ecological destruction. Despite occasional statements from various representatives of communist countries that they have neither poverty nor pollution in their countries, the end of the Cold War peeled back the curtain to reveal both poverty and environmental damage [8].

Funding for any UN endeavor rests upon finding Northern country support, not only for the meeting but also for the participation of those states that would otherwise not be able to participate. According to Luchins [19], cost factors played a role at the beginning of the deliberations to determine the scope of the negotiating session. After this point, occasional concerns about costs continue to be raised by the developed countries. Luchins speculates that Southern states may well have used cost concerns as a decoy, possibly due to France's insistence upon continuing nuclear testing despite mounting evidence of widespread damage to the French Polynesia territory, in an attempt to set up a quid pro quo trade on these issues.

As with any major diplomatic endeavor, the time frame for the formal conference occurs after a series of lengthy and intensive negotiating sessions. These meetings begin by establishing rules by which states agree to be bound during the negotiations. During PrepComs, states define actual treaty text and identify areas where states agree and disagree. Consequently, conference organizers spend a significant amount of organizational time and effort into creating successful PrepComs, including conducting diplomatic negotiations.

For UNCHE, states agreed to limit membership on the PrepCom to 27 countries. The number of states participating in a PrepCom varies, but the UN strives to maintain a regional balance during negotiations and when hosting major negotiating events. Accordingly, the first PrepCom session occurred from March 10–20, 1970. Early conversations during this meeting primarily involved the Northern industrialized countries as the Southern position viewed the entire international environmental agenda as a ruse to force newly created states in Africa and Asia back into a subservient colonial role [12]. Thus, the entire conference was not worthy of Southern support or participation.

Brazil, in particular, voiced grave concerns about the Northern motivations for proposing limiting environmental pollution. In short, some Southern states were willing to trade pollution for industrialization and the potential to generate wealth [12]. Consequently, the lack of Southern support weakened the ability to negotiate a stronger diplomatic outcome and limited environmental protection to expressing principles rather than proposing limits to environmentally damaging activities.

That is not to say that the Southern bloc displayed a homogeneous negotiating position. China, which recently gained membership within the UN, did not play an impactful role leading up to the conference because of its unfamiliarity with the UN system. However, they, along with Brazil, articulated the "additionality" concept [12]. Northern states who had industrialized were responsible for cleaning up the environment within their borders. Further, the Northern countries should also pay for and were also responsible for environmental remediation efforts in the developing countries as the instigator and benefactor of the pollution. As used during this time frame, additionality went further than mere environmental remediation efforts. However, this conceptualization also included the opinion that developed countries that created an environmental standard as a threshold requirement for importing goods should be responsible for any financial harm to developing countries in order to keep the Southern countries' economic interests free from financial harm.

Interestingly, this issue did not remain within the PrepCom but engulfed the UN GA. Since the G-77 countries held the majority in the UN GA, Southern concerns would not only be taken into consideration but would come to dominate and define the entire environmental agenda. Once brought to the UN GA's attention, Southern countries passed a resolution in 1970 stipulating the addition of development concerns to the conference agenda [29]. This addition proved timely as the momentum to draft the conference outcomes strengthened at the second PrepCom held February 8–19,

1971. Working drafts of the two Stockholm Declaration of Principles began at this time and continued into the third PrepCom held September 13–24, 1971.

In addition to the PrepCom process, Secretary-General of the UNCHE, Maurice Strong of Canada, convened additional conferences to continue the conversation about the interlinkages between environment and development. Strong was not the first choice to lead the UNCHE process. Jean Mussard served as the original director of the Conference secretariat until autumn 1970. Strong, a Canadian businessman, brought personal leadership and the ability to generate trust and cooperation with him. Using the slogan "Only One Earth," Strong charmed an increasingly broad and diverse coalition including states, private organizations, and influential thought leaders to support the conference's goals with an emphasis on the need to reconcile the two very different perceptions of the relationship between environment and development.

In an interesting departure from the norms of international diplomacy, Strong had no qualms about engaging resources outside of the state system to build political support and momentum. One of the most significant of these other groups proved to be the Founex Conference, held in Founex, Switzerland, June 4–12, 1971. Arrangements for the conference benefitted from the engagement of renowned development experts Mahbub ul Huq, Gamani Corea, and Barbara Ward who articulated the principle of eco-development. The term means that financial assistance and technology transfer should be encouraged to avoid traveling the same development path that led to the current state of environmental damage. While this conference may not seem important compared to the size and scope of the UNCHE, this conference marks the beginning of articulating sustainable development, one of the leading paradigms in international environmental affairs today.

In the meantime, Southern countries utilized the 1971 UN GA to further force the conference agenda toward the Southern preferences for the intertwining of environmental and developmental issue areas. UN GA Resolution 2849 (XXVI) of December 20, 1971, further aggravated North–South tensions by calling on the Northern countries to finance any new pollution controls necessary for all Southern countries. Unsurprisingly, none of the Northern countries voted for this resolution, opting instead to either vote against the resolution or abstain from the vote.

Neither the North–South gap nor the Soviet boycott prevented work on the Stockholm Action Plan or the Stockholm Declaration during the fourth PrepCom held March 6–10, 1972. The Stockholm Action Plan recommended concrete actions that the UN agencies and its member states could take to foster communication and cooperation. The Stockholm Declaration outlines a series of soft law principles, guidelines that member states should follow but do not amount to a legally binding treaty. Much of international environmental law takes the form of a code of conduct rather than a law or regulation with fines or jail times for failing to adhere to the principles. Instead, states are expected to act in accordance with treaty guidelines that may, over time, become customary international law.

Additionally, PrepCom participants considered plans to form the UNEP after the conference's conclusions. Von Molke [30] took a cynical view of the action to create what would become the UNEP by pointing out that no existing specialized agency within the UN system wanted to take responsibility for this issue. The United States supported the creation of UNEP [12]. Additional ideas to create a UN Environment Fund also aired during this meeting. Southern opposition emerged, however, in that the United States proposal allowed Northern countries too much control, and conversations on how to fund and structure UNEP continued throughout the period leading up to the Stockholm session.

Away from the PrepCom process, Strong engaged all elements of civil society to build support for the meeting. Strong recruited a group of scientists and development experts to research on behalf of the conference. Barbara Ward became one of the leading figures within this group. Along with René Dubos, a microbiologist, she authored *Only One Earth*, a standalone manifesto that incorporated the philosophy of limited capital resources with the Southern perspective on development. This officially unofficial book sought input from numerous others and simultaneously strengthened the scientific community's standing and the increasingly relevant environmental NGOs' support for the UNCHE process. Ward and Dubos [31] not only defined the agenda for the diplomats about to convene in Stockholm, they also articulated the rationale for cooperation to stave off certain environmental doom.

Ward, a British economist and Schweitzer Professor of Economic Development at Colombia University in New York City, possessed an influential voice that recognized the need for states to operate within limits created by the ecological system. As part of the preparation for Stockholm, she became one of the leading voices for Southern concerns, including establishing the International Institute for Environment and Development (IIED) in 1971. Indeed, Ward's contributions to UNCHE were so significant that delegates suspended the official meeting so that she could address the plenary. Ward's efforts on environmental and development issues extended well beyond the end of the UNCHE meeting. Queen Elizabeth named her a life peer after she retired from teaching by awarding her the title of Baroness Jackson of Lodsworth in 1976.

2.3   At Stockholm

Today, history smiles favorably upon this first foray into international environmental politics, but the productive outcome of the conference during its conception was anything but assured. The institutional infrastructure that many scholars, students, and practitioners would expect at a major global conference remained in the process of becoming established. Many countries, including the United States of America, were creating their domestic environmental ministry or equivalent federal agency.

Conference secretariat Maurice Strong skillfully negotiated the intractable and divisive ideologies that threatened the conference. In the process, he launched the search for new principles with the ability to reconcile the differences between East and West and between the North and South perspectives on environmental damage. While Strong was unable to overcome the great power dilemma that prevented East Germany from attending the conference, the expected drop in legitimacy stemming from the Soviet boycott never materialized. The Soviet Union participated in the PrepCom negotiation almost to the end. Additionally, Strong successfully added Soviet citizens, most notably, the Soviet Academy of Sciences member Vladimir Kunin, to his staff in his personal capacity. Further, Soviet Union diplomats traveled to Stockholm during the UNCHE conference but never formally appeared at the meeting. This strategic move undoubtedly allowed the Soviet Union to stay informed about the conference despite their boycott of the fourth PrepCom and the Conference itself.

The North–South gap also showed no signs of abating. Indeed, the North–South gap remains one of the signature divisions within international environmental politics. However, Strong's background with international aid enhanced his ability to navigate a path forward between the two sides [12]. The clever articulation of eco-development allowed enough of a victory for both sides to move forward without reducing momentum toward enhanced cooperation to limit environmental pollution or negating Southern desires for increased financial aid in the quest to develop in an environmentally sensitive direction.

Head [17] summarized the conference as a carefully constructed compromise that allowed Southern countries to continue to develop, but with the understanding that the developing countries would take steps to minimize environmental damage. In return, the Northern countries donated development assistance to help the South avoid some of the more damaging environmental problems the North encountered during industrialization. Further, the North also agreed to lower its environmental pollution in an attempt to protect both domestic citizens and to decrease the health impacts on the receiving country.

The North's negotiating position was set forth by the so-called Brussels group consisting of Belgium, France, Germany, Italy, and the United Kingdom. These countries did not necessarily support the calls for firm limits on pollution or a powerful environmental organization. While the United States and the Netherlands also appeared as members of this group, both countries were more supportive of the Stockholm outcomes [12].

During Stockholm itself, the stance of the Southern countries noticeably softened since the walkout that these countries threatened during the run-up to the conference did not materialize. Strong's determination to incorporate Southern concerns into the meeting agenda prevented the walkout. He also found an able ally in India's Prime Minister, Indira Gandhi. Her engagement at Stockholm helped reassure the other developing countries that environmental measures would not harm Southern interests. Further, she ably championed the linkages between environment and development as a necessity to lift people out of poverty.

By the end of the Conference, states finalized two notable documents and successfully established an outline for future environmental discussions. The Stockholm outcomes included the Declaration of Principles, some of which have entered into customary international law. The second document, the Stockholm Action Plan, elucidated 109 recommendations that included both financial and institutional arrangements.

Within the Stockholm Declaration of Principles, diplomats intended, in Principle 21, to strike a balance between national sovereignty and the good neighbor ideal [25]. Consequently, while states are sovereign within their national jurisdictions, the state nonetheless maintains responsibility for limiting the impacts of pollution on other states. Many people believe this principle is sufficiently well established to have entered into customary international environmental law [15]. States may be bound to follow customary international law, regardless of whether the state has embedded the principle into domestic legislation or signed a treaty formally accepting the principle. This is because the international diplomatic system interprets silence as consent.

The Stockholm Action Plan called for the creation of UNEP that heralded the permanent addition of environmental and developmental ideals to the UN agenda. Important recommendations of note included items on human settlements (Recommendation 3), protecting species within international waters (Recommendation 32), and a 10-year moratorium on whaling (Recommendation 33). Perhaps the most far-reaching addition to the international architecture occurred when Recommendation 14 suggested the need for a new intergovernmental body working on environmental affairs. While not often discussed, the Stockholm Action Plan also contained an outline that environmental negotiations should engage in environmental assessment, followed by environmental management. The environmental assessment stage took into consideration the status of the environment. Environmental management, according to Fritz [14], refers to managing the international system to work toward environmental protection. Management of the international system can conflict with state sovereignty that reserved the management of the domestic environment exclusively for the appropriate national government. International environmental diplomacy also included the supporting measures necessary to ensure that the international organization would have the resources needed to carry out the tasks that states assigned. Further, many developing countries would need financial and technical assistance to meet their treaty obligations.

The Stockholm Action Plan did not mean that states alone championed the cause of environmental protection. Other actors, now commonly referred to as the non-state actors, also arrived in Stockholm with the intent to influence the conference outcomes. This catch-all category of groups that occasionally interact with international diplomacy includes scientists, NGOs, business and industry groups, and the media.

Regardless of their official status, Conference Secretary-General Strong possessed the charisma that reassured the plethora of voices seeking that they would have the opportunity to influence the conference. More importantly, Strong followed through this verbal commitment with actions that granted participants some role in the conference proceedings, even if that role was not as prominent as the member state's role. For example, Strong arranged that biologists sponsored by Dai Dong participated by arranging a meeting with then Secretary-General U Thant to accept a treatise on the role of science in shaping the environmental agenda [27].

Perhaps one of the underrated achievements of the Stockholm conference involved Strong's subtle molding of the institutional arrangements to allow non-state actors to give scientific and political advice. NGOs certainly impacted the city of Stockholm during the conference as they engaged with media and other interested parties in the official Environment Forum and at the three unofficial conferences, including the Dai Dong conference, the Folkets Forum, and the Life Farm. Strong enthusiastically encouraged interested attendees to participate in one of three side event conferences.

The UN officially sponsored the Environment Forum, and this gave NGOs accredited to this forum limited access to conference delegates and more limited ability to address the conference directly. However, the strong leadership of this group faltered, and Barry Commoner, an American biologist and one of the leaders of the environmental movement, took over the event to rail against US policies, in particular the VietNam war. Ironically, the Forum quickly gathered a reputation for an undiplomatic veneer, including a lack of respect for free speech and occasional forays into actual violence [1, 16, 26].

In contrast, the Dai Dong conference held immediately before the Stockholm conference suffered from none of the mismanagement or misbehavior of the Environment Forum. The conference organizers set out to produce a draft treaty text and a concrete action plan utilizing the Southern perspective. While the conference generally accomplished this goal, the participants failed to achieve the consensus of the main conference, with six scientists withholding signature until a statement that repudiated some of the content had been added. Rowland [22] indicated that UN diplomats finalized agreements on more contentious issues such as population, while participants of the Dai Dong conference failed to achieve consensus. However, this is perhaps understandable as the Dai Dong conference participants did not have quite the same set of pressures to produce a statement as the diplomats gathered under the UN banner.

The Folkets Forum or People's Forum represented the views of left-wing environmental and political groups. It adapted admittedly Marxist philosophies on the relationship between limits to growth, population, and equal distribution of wealth that would not have been acceptable to mainstream thought in the Northern countries, especially Western Europe and the United States. While the Folkets Forum may have been the best organized of the unofficial conferences, there is scant evidence that this conference significantly impacted the delegates or the Stockholm outcomes.

In contrast to the scientific respectability of the Dai Dong conference, the Life Forum, or Hog Farm as the media more commonly named it, represented vibrancy. Headed loosely by the Merry Pranksters, the Life Forum was best represented by the paper mâché whale that the Merry Pranksters drove around Stockholm for the conference duration. This eclectic piece of street theater provided international media outlets with a colorful view perfect for the nightly news, along with a quick "save the whales" soundbite. This mobilization of media attention increased the likelihood that the official documents would include an action item on whaling. It most certainly vaulted the whale into prominence as a symbol of international environmental conservation.

Evidence suggests that NGOs did have some modicum of success when speaking to delegates. Eco [10] story "NGO come back, then go home" praised their contributions to the discussion after hearing Margaret Mead speak on behalf of a coalition of NGOs. While details of the draft treaty text presented by the NGOs are not readily available, Artin's [1] second-hand account suggests that he spoke about both substantive and procedural issues. Mead [21], reflecting on the occasion during a different speech given before the Governing Council of UNEP, provided further details when she stated that Barbara Ward, the author of *Only One Earth*, wrote the Declaration of the NGOs with input from over 150 groups in attendance at the conference. Interestingly, the coalition named by Artin and described by Mead included both environmental NGOs and business and industry NGOs. Given the newness of the issue area and the possibility of many first-time attendees at a UN conference, NGOs faced their own learning curve on interacting with the UN. Consequently, NGOs agreed to build coalitions to interface with the UN more effectively [11].

If direct evidence on environmental NGOs' actions at this milestone conference appears scarce, first-hand information on business and industry groups remains more challenging to come by. Perhaps more is known about business and industry groups from the critiques leveled by environmental groups than from positive accounts generated by businesses themselves. For example, Artin [1] claimed that the Stockholm conference gave impetus to Sweden's blatant promotion of goods and services that negatively impacted the environment. He was particularly scornful of the welcoming packet that the Swedish government produced for official delegates that included vinyl cases, advertising slicks that promoted further industrialization and urbanization that increases pollution, and the promotion of Swedish industries, including car manufacturer Volvo. However, no information to date has been located involving the negotiating positions of these groups.

A number of NGOs were attempting to influence the conference outcomes through various formal and informal mechanisms. One of these mechanisms included the *Eco* bulletin, a daily news brief, that was produced for the duration of the Stockholm Conference. Its self-appointed task was to report accurately on the negotiations, albeit with a writer's bias that favored pro-environmental positions. Interestingly, *Eco* accepted ads from industries promoting special tours and advertising their nascent green credentials. Additionally, it laid the groundwork for future specialty publications.

When taken into consideration, the Stockholm Conference created the principles that would initially dominate the international environmental agenda, it also established expectations for future relationships. Defining characteristics that reappear include the North–South gap, the role of NGOs and industry experts as technical and political consultants, and the process by which states would attempt to solve environmental problems.

2.4   After Stockholm, 1972

The beginning of this chapter identified several environmental problems that led to the Stockholm Conference, including pesticide use and oil spills. In the run-up to Stockholm, other environmental issues emerged, such as water pollution, acid rain, toxic air pollution, and lost or endangered species, among other problems. It is important to note that the Stockholm Action Plan failed to implement potential solutions to these environmental problems. However, this does not mean that the Stockholm Conference should be thought of as a failure for a variety of reasons. UNCHE made a lasting positive contribution toward environmental protection because it established a useable framework that continues to function within the international diplomatic system.

First, UNCHE established the process through which environmental cooperation could be negotiated, and that pattern remains in existence some fifty years later. This pattern includes the invitation of states to meet to discuss a specific environmental problem. This meeting is then "observed" by NGOs representing various special interests, including business and industry groups, environmental organizations, labor unions, religious groups, and any other subset willing to organize and support UN activity. Additionally, scientists formed consultative relationships with various actors, including states, the UN, and other NGOs.

Much conversation has been made scholastically about the Stockholm-Rio-Johannesburg mega-conference trajectory within international politics. Scholarly works ascribed six functions to the mega-conferences: determining a global agenda, promoting environmental and developmental linkages, endorsing common principles, providing global leadership, building institutional capacity, and creating support for the processes and outcomes by promoting inclusivity [23, 24].

Second, UNCHE recommended the establishment of the UNEP with its focus on the linkages between industrialization and environmental pollution and the linkages between poverty and pollution. While UNCHE did not have the authority to officially begin UNEP, given that the same states that comprise the UN GA also met as part of the UNCHE conference, the establishment of UNEP at Stockholm may be considered a fait accompli. Diplomats at Stockholm cleverly worked around the fact that other specialized agencies worked, in practice, on environmental affairs and had accumulated significant expertise on the subject. Thus, diplomats declared that UNEP would be designed utilizing the "form follows function" principle, meaning that UNEP would be designed to work on those items that other specialized agencies did not. Ivanova [18] argued that the normative and catalytic functions demanded of UNEP plugged a gap in the institutional infrastructure of the UN without infringing upon other specialized agencies that already had specific mandates such as the World Health Organization (WHO) or the World Meteorological Organization (WMO).

UNEP's primary function within the international system is to encourage a greater understanding of environmental problems and catalyze actions that protect the environment. UNEP serves as a visible, vocal champion of the environment within the international system. That being said, scholars do not consider UNEP a strong environmental organization, particularly when compared to the United Nations Development Program (UNDP) and calls for creating a new world environmental organization appear periodically [3].

Third, Stockholm expanded the types of actors able to access formal negotiating sessions, and this, in turn, altered the agency of the international system. This process added environmental and development affairs to the numerous issues covered by international diplomacy and expanded the actors actively participating in conference negotiations. At the height of the Cold War, the international system focused almost exclusively on the power of the state within a bipolar state system. States typically did not acknowledge any non-state entity and the Stockholm conference with its NGOs "go home" attitude reflected this ideal. Similarly, neither scientists nor business and industry groups received any type of special attention, despite the fact that both groups actively engaged in both the PrepCom process and at the Stockholm convention itself.

Scholarship at the time reflected this mindset. The international system operates as the exclusive role of sovereign national governments. Each government is legally equal, regardless of the differences in wealth, technology, population, military prowess, knowledge, or skill at diplomacy. International environmental affairs, then, should be negotiated through the states who would then oversee the implementation of agreements, each in their own country. Population control, combined with new pollution abatement equipment, would suffice, for now, as an attempt to restore the environment for that group of Northern countries that saw industrialization as the cause of environmental damage. Southern countries that framed environmental damage as a function of poverty hoped instead for an increase in economic wealth that not only would ensure their continued independence but allow them to compete more equally with the North in the future.

It is clear from this model that few people saw a permanent role for the non-state actors within international environmental politics. At best, environmental NGOs would serve as unofficial technical advisors [13]. It was not clear at this time whether the role of environmental NGOs would carry over after the end of the conference. Egelston [11] believes that NGOs in Stockholm sought not only to influence the outcomes of UNCHE but also to gain access to UN decision-makers over the long run. NGOs would exchange technical knowledge about environmental damage and knowledge about solutions the general public would accept for a consultative arrangement with UN civil service employees. This arrangement would allow NGOs to shape future political agendas of a multitude of other actors, including sovereign states, other international organizations, other NGOs, and business and industry members.

After Stockholm, environmental affairs gained a permanent place on the global agenda. While influential scientific voices such as Barry Commoner critiqued the meeting for not banning environmentally polluting industrial processes, the creation of new norms to protect the environment, and the intent to begin a new specialized agency focused on these concerns advanced environmental protection [4]. Whether the world's countries could remain united in the face of contested differences and economic divides remained to be seen. Equally important, the conceptualization of the sovereign state as the only representative of the territory and, therefore, an unchallenged authority on environmental problems would also become contested in the future.

References

1. Artin T (1973) Earth talk independent voices on the environment. Grossman Publishers, New York
2. Bell ML, Davis DL, Fletcher T (2004) A retrospective assessment of mortality from the London smog episode of 1952: the role of influenza and pollution. Environ Health Persp 112(1):6–8

3. Biermann F, Bauer S (2017) A world environment organization: solution or threat for effective international environmental governance? Routledge, New York
4. Black R (2012, June 4) Stockholm: birth of a green generation. BBC News. Retrieved from https://www.bbc.com/news/science-environment-18315205
5. Brimblecombe P (1976) Attitudes and responses towards air pollution in medieval England. J Air Control Assoc 26(10):941–994
6. Carson R (1962) Silent spring. Houghton Mifflin, Boston
7. Chertow MR (2000) The IPAT equation and its variants. J Ind Ecol 4(4):13–29
8. Dominick R (1998) Capitalism, communism, and environmental protection: lessons from the German experience. Environ Hist-US 3(3):311–332
9. ECOSOC (1968) Consideration of the provisional agenda for the forty-fifth session. E/4466/Add.1
10. Eco (1972) NGO come back, then go home. Stockholm: 14 June 1972 Bound as Stockholm Conference Eco, vol 1
11. Egelston AE (2012) Sustainable development: a history. Springer, New York
12. Engfeldt L-G (1973) The United Nations and the human environment - some experiences. Int Organ 27(3):393–412
13. Feraru AT (1974) Transnational political interests and the global environment. Int Organ 28(1):31–60
14. Fritz J-S (1997) Earthwatch 25 years on: between science and international environmental governance. IIASA Interim Report. IIASA, Laxenburg, Austria: IR-97–059
15. Harlow BA, McGregor ME (1996) International environmental law considerations during military operations other than war. Intl Law Stud 69:315–332
16. Hawkes N (1972) Stockholm: politicking, confusion, but some agreements reached. Science 176(4041):1308–1310
17. Head JW (1978) Challenge of international environmental management: critique of the United Nations Environment Programme. Va J Int Law 18(2):269–288
18. Ivanova M (2007) Designing the United Nations Environment Programme: a story of compromise and confrontation. Int Environ Agreem-P 7(4):337–361
19. Luchins D (1977) The United Nations Conference on the human environment a case study of emerging political alignments. Dissertation, City University of New York
20. Malthus TR (1986) An essay on the principle of population 1798. The works of Thomas Robert Malthus. Pickering & Chatto Publishers, London
21. Mead M (1977) Human settlements and the UN Environment Program: The potential of the Non-Governmental Organizations (NGOs). Ekistics 44(262):177–183
22. Rowland W (1973) The plot to save the world. Clarke, Irwin & Company Limited, Toronto
23. Seyfang G (2003) Environmental mega-conferences—from Stockholm to Johannesburg and beyond. Global Environ Chang 13(3):223–228
24. Seyfang G, Jordan A (2002) "Mega" environmental conferences: vehicles for effective, long term environmental planning? In: Stokke S, Thommesen O (eds) Yearbook of international cooperation on environment and development. Earthscan, London, pp 19–26
25. Sohn LB (1973) Stockholm declaration on the human environment. Harvard Int Law J 14(3):423–515
26. Stone PB (1973) Did we save the earth at Stockholm? Earth Island Limited, London
27. UNESCO (1971) "S.O.S. save our environment" UNESCO Courier, UNESCO, Paris, p 4–5
28. UN GA (1969). Twenty fourth session official records second committee 1276th meeting
29. UN GA (1970). United Nations Conference on the Human Environment. A/RES/2657(XXV)
30. Von Moltke K (1996) Why UNEP matters. In: Bergesen HO, Parmann G (eds) Green globe yearbook of international co-operation on environment and development 1996: an independent publication on environment and development from the Fridtjof Nansen Institute. Oxford University Press, Norway, pp 55–64
31. Ward B, Dubos R (1972) Only one earth: the care and maintenance of a small planet. Penguin Books Ltd., Harmondsworth

Chapter 3
Institutionalizing UNEP

Abstract At the end of the UNCHE in 1972, the street fair atmosphere created by the convergence of the international environmental community in Stockholm, Sweden, ended when its participants returned to their respective countries. However, the creative energy that spurred cooperation on behalf of the environment did not evaporate. Thus, the forward momentum and goodwill to implement the Stockholm Action Plan dispersed around the world, heading back into the hallways of the existing UN work locations, most notably the UN Headquarters in New York, and into various capital cities around the globe. This chapter reviews UNEP's creation and reviews its primary functions of environmental assessment, environmental monitoring, and catalyzing cooperation. It also introduces readers to a concise overview of realism and liberalism, two international relations theories sometimes utilized to understand international environmental diplomacy. As many environmental proponents emphasize cooperation over conflict, liberalist models may be preferred in this subfield.

Keywords United Nations Environment Program · United Nations Conference on the Human Environment · Stockholm Action Plan · Earthwatch · Realism · Liberalism

At the end of the UNCHE in 1972, the street fair atmosphere created by the convergence of the international environmental community in Stockholm, Sweden, ended when its participants returned to their respective countries. However, the creative energy that spurred cooperation on behalf of the environment did not evaporate. As stated in the Stockholm Action Plan, "What is needed is an enthusiastic but calm state of mind and intense but orderly work." ([34]: 3). Thus, the forward momentum and goodwill to implement the Stockholm Action Plan dispersed around the world, heading back into the hallways of the existing UN work locations, most notably the UN Headquarters in New York, and into various capital cities around the globe.

This intensive work to create new international infrastructure focused on environmental affairs utilizing the Stockholm Action Plan as the blueprint for the future envisioned by the conference participants. However, once the hard work of implementing the Stockholm Action Plan began, significant divides within the international community reappeared. Japan, for instance, deeply resented the moratorium on whaling that occurred in Recommendation 33 and saw the ban as a direct assault on the Japanese culture as whale meat remains an import part of food consumption.

While Japan's reaction may have been premature as states did not finalize the ban on whaling during the conference, the inclusion of a recommendation within the Stockholm Action Plan nevertheless signaled intent, even if this document did not constitute final action. Thus, the Stockholm Declaration outlined new normative principles necessary to protect the environment.

The adoption of these documents, while a significant step forward, did not represent a legally binding requirement to complete the recommended changes. For example, the Stockholm Action plan envisioned a new environmental organization, UNEP. Diplomats writing the document perhaps did not foresee an organization located in Nairobi, Kenya, with a weak mandate for environmental change dependent on voluntary funding for its activities.

While the creation of UNEP occurred swiftly upon the conclusion of the UNCHE meeting, states changing behaviors to adopt these new environmental norms remains a debatable topic at this time. Adopting new norms as implied in the Stockholm Declaration frequently represents a lengthy process. Very few, if any, scholars would argue in the aftermath of the Stockholm Conference that the international system underwent an immediate transformation to incorporate environmental concerns fully into political or economic decision-making.

This chapter, then, investigates the implementation of the Stockholm outcomes by examining four central themes. Section one focuses on creating UNEP by examining critical issues related to its location, organizational structure, and funding. This section also discusses UNEP's relationships with other portions of the international system, including other UN specialized agencies, to assess the extent to which UNEP's efforts reflected the will of states within international diplomacy. This section also introduces classical realism as a fundamental international relations theory by which students and participants may analyze events to determine when states successfully utilized their political power and influence to achieve their goals.

Section two examines the tasks assigned to UNEP, namely, to assess current environmental quality and assist states in managing the environment, including monitoring essential information about scientific updates and technological progress. These two items mesh because successfully managing environmental quality depends on relevant scientific information. Further, UNEP's ability to complete this task relied and continues to rely on outside sources with scientific and technological knowledge that various segments of the global society may contest. Given the scope and complexity of the task at hand, UNEP depends, by necessity, on assistance from other key actors to complete these tasks. Thus, this section also looks at why states and individuals cooperate via liberalist international relations theory.

Section three looks at early attempts to create new norms and principles around the environment and development's key themes. This section takes an in-depth look at the Stockholm Declaration to examine the existing norms around state behaviors, the dominant economic growth paradigm, and the emerging environmental norms

proposed within this document. A third international relations theory, constructivism, focuses on how changes in the social reality in which the international state
system operates may also explain how this system changes over time. Section four
concludes the chapter by critiquing UNEP's activities during this time frame. The
section focuses on scholarly thought on UNEP by noting that academic opinion
on this organization remains divided. This division may reflect different theoretical
models of international relations in use to determine the effectiveness of international
organizations.

3.1 Establishing UNEP

Given the amount of work contained within the Stockholm Declaration and the Action
Plan, international diplomats and civil servants began institutionalizing environmental affairs within the UN system; diplomats began creating a new program within
the UN. Within the international state system, the meaning of structure depends, in
part, on the theory utilized to model international relations [38]. The international
structure encompasses more than the formal rules by which an international organization operates. That is, the structure incorporates not only the "rules of the game"
but also the relationships that both promote and constrain each actor as they work
toward obtaining their preferred outcomes, including the normative ideas, values,
and beliefs that each agent promotes.

On December 15, 1972, the UN GA passed Resolution 2997 (XXVII), establishing
UNEP. Totaling a mere three pages, the Resolution formalized the agreements made
at Stockholm. Internally, UNEP consisted of a fifty-eight-member Governing Council
with representation allotted on a regional basis (consistent with UN practices in other
specialized agencies and programs), an executive director, and a small secretariat.
The UN GA directs UNEP's activities through the Governing Council. As the UN GA
is itself dominated by developing states, this reporting structure assured that UNEPs
activities conform to the Southern states' desires while ensuring some representation
from the Northern states due to the regional composition rule. The Liaison Officer for
Sweden at the time of the UNCHE, Engfeldt [7], reports that the original composition
of the Governing Council sidestepped the question of which Germany should be
recognized internationally by awarding seats to both countries. Thus, the Governing
Council cleverly avoided becoming embroiled immediately in one of the Cold War
diplomatic conflicts.

In addition to UN GA oversight, UNEP also generates reports for the ECOSOC,
meaning that ECOSOC will also have the ability to direct the work of UNEP.
Established as one of the six main organs of the UN, ECOSOC directs much of
the work within the UN system as it coordinates a variety of specialized agencies, including humanitarian crises, emerging issues, developmental policy, and
environmental affairs.

The UN GA elects UNEP's executive director. This position not only manages
the day-to-day operations but has considerable autonomy, including the decisions to

suggest strategic directions and approve funding individual projects via the newly created Environmental Fund. Thus, this position will be functionally independent of the Governing Council and able to bring up issues that the Governing Council may not wish to consider. Additionally, this position is intended to be independent and authoritative, with a strong ability to interface directly with the other specialized agencies within the UN system. Unsurprisingly, states preferred Maurice Strong for UNEP's first executive director.

UNEP's primary function within the international system is to encourage a greater understanding of environmental problems and catalyze actions that protect the environment. However, what exactly is meant by an environmental problem differs. Hardy [13] pointed out that whether the environment also included development concerns such as habitat or human settlements remained uncertain throughout this time. He also noted that the second line of the debate focused on the activities UNEP would be best positioned to accomplish, given that many other UN entities also worked on these issues.

UNEP's role included reviewing the implementation of international environmental actions, advising states and other UN agencies on creating and implementing new programs, and integrating knowledge from scientists and other relevant professional communities into this diplomatic decision-making process. Notably, these functions do not grant UNEP the power to implement or enforce environmental programs directly. Indeed, the language considered at PrepCom III in the run-up to the Stockholm Conference suggested that UNEP should not compete with the agency it is attempting to influence [33]. This institutional setup preserves the piecemeal approach to environmental affairs already present in the UN system to a certain extent. The United Nations Educational, Scientific and Cultural Organization (UNESCO), the sponsor of the 1968 Man and the Biosphere Conference in 1968, expressed grave concern that the creation of UNEP would subsume its work [7].

The pre-existing programs instituted by other portions of the UN did not transfer to the newly created UNEP. For example, the WHO, headquartered in Geneva, Switzerland, worked on increasing the quality of sanitation systems as well as encouraging the adoption of new designs as a potential solution for eliminating the causes of diseases, an item of vital importance to the developing countries. Additionally, the FAO, headquartered in Rome, Italy, finished a four-year blueprint for increasing agricultural development in 1969. The Indicative World Plan for Agriculture Development included recommendations for land and water use. Proponents for improving environmental quality could easily conflict with these projects, especially on items concerning the use of fertilizers and pesticides, preventing agricultural runoff into waterways, and converting forests into agricultural land use.

The new environmental organization would also interface with the UNDP, headquartered in New York. Recently created in 1965 through the combination of the UN Expanded Program of Technical Assistance and the UN Special Fund, UNDP sought to assist countries in building the social and economic infrastructure needed to improve the quality of life for its citizens. Sample projects included increasing agricultural production, city planning and development, and providing funds for economic expansion. These activities could generate new environmental damage, creating the potential for conflict between the two programs. Thus, the newly minted UNEP would need to establish an operational vision that did not infringe upon the mission of already established specialized agencies and programs within the UN system.

In addition to establishing a unique role for UNEP, states also needed to secure funding for the program. Southern countries articulated that this funding should be added to other international development assistance funds, increasing the overall aid level. Northern countries, however, would be the main contributor to funding UNEP. This, in turn, raises an important consideration about the amount of financing the Northern states should provide. Concerns emerged that the Northern countries, the primary source of industrialized pollution, would use the funding to control and limit UNEP's activities and mute potential future criticisms. Further, UNEP could be hampered from the onset by powerful coalitions that did not care if UNEP succeeded; or, worse, wanted UNEP to fail.

While states should communicate an interest in a new international environmental organization, it is more significant when states follow through on that interest by formally creating a new agency. This requires a permanent funding commitment for buildings, personnel, equipment, and travel. The absence of friendly interests willing and able to protect the fledgling program would allow UNEP to exist but also allows states to disregard the organization. However, openly hostile countries could potentially be fatal to the program and the planet. Resolving these differences left UNEP financially constrained as funding for UNEP occurs in a piecemeal approach in that salaries and expenses for the Governing Council, executive director, and secretariat come from the regular UNEP budget. Other programmatic funding comes from voluntary contributions, with a heavy reliance on funding from the United States, which assumed responsibility for providing 40% of the initial $100 million for the first five years of the Environmental Fund [34].

In a symbolic victory for the Southern countries, the UN system selected Nairobi, Kenya, as the headquarters for UNEP, making it the only major subunit headquartered in a developing country. While states considered locating UNEP near other central operating units in Europe, the overall sentiment supported the symbolism of locating a new organ in the developing South. Kenya secured this headquarters as this country had become the leader of the African countries, including winning a coveted seat on the UN Security Council, partly due to its strong support for anti-colonialism, the dominant paradigm for Southern countries during this time frame [39]. A more cynical view of this move alleges that the developed countries supported this move to the South to isolate UNEP so that it would not be able to tamper with economic wealth and development action items. In hindsight, the placement of UNEP in Nairobi simultaneously weakened UNEP's ability to influence other UN specialized agencies but probably allowed Southern countries greater access to shape the future of this agency.

As part of the UN system, UNEP answers to its member states; thus, the model created during UNCHE presumed that UNEP would provide scientific advice to states who would then act based upon this information. Similarly, UNEP's efforts to promote environmental consciousness would also encourage states to work toward mitigating or eliminating environmental threats that crossed national boundaries.

Consequently, this organizational structure firmly limited UNEPs' activities. Most importantly, states retained their privileged position in the international system as the only legitimate implementer of environmental activities. UNEP does not have the authority to force a state to take any action the state does not want to take. States can and do cause environmental damage. The United States government is one of the largest polluters globally due to its use of fossil fuels, especially in the United States military [4]. Equally important, states also have sovereignty over their businesses and industries. States may choose to protect these economic interests at the expense of a cleaner environment.

Also noteworthy, UNEP's institutional design sets member states above the office of executive director. Thus, while the executive director enjoys some significant autonomy, the executive director's office ultimately reports to the Governing Council. Because the Governing Council consists of member states, this structure places these states who may themselves be a significant source of environmental damage over the agency charged with catalyzing activities to reduce that very same environmental damage.

While much of the scholarly work assessing Stockholm focused on how states expanded the UN agenda, very little attention has focused on the role of the non-state actor in the aftermath of the Stockholm conference. Egelston [6] provided a brief glimpse of the role of non-state actors when she describes that non-state actors focused on building relationships with the newly created UNEP and with other NGOs. UNEP's mandate to collect information about the amount and causes of environmental damage around the globe meant that it needed a collection structure outside of the control of states who might not want to admit to the poor environmental management domestically. Thus, environmental NGOs served as both paid and unpaid consultants to UNEP.

Environmental politics demonstrates some of the traits associated with realist international relations theory. The realist view of the world emphasizes the roles and actions of states within environmental politics. Classical realism believed that states seek power, with the ultimate power as the ability to be free from the political dictates of others [27]. Frequently, scholarship within international relations determines that power must therefore be cast as military superiority or economic wealth. Thus, states act in their self-interest to acquire military power and financial wealth as a means of exerting power over others. The ultimate success within realist theory is to become a hegemon, one of the most powerful actors, if not the most powerful actor, within the international state system. Consequently, states struggle against each other in perpetual conflict.

Powerful states dominate not only weaker states but also international organizations. States with significant power may work through an international organization rather than acting outright [1]. Thus, states and international organizations become interdependent on each other. International organizations provide an efficient method of managing conflict and cooperation between states. States, in turn, benefit from the neutrality of the international organization, especially in situations involving disputes.

This theory adequately explains the wishes of the Southern states to be free of their colonialist past, including the vigor with which Southern states defended their right to self-determination and economic growth. Realism theory dictates that these new states acquire power through creating new industries that generate economic wealth regardless of the impact on the local environment.

Realist theory also supports the idea that UNEP's founding occurred, in some significant part, because the United States wanted it to happen. As the leader of the "Free West," the United States provided the military power, economic wealth, and political leadership that promoted democratic ideals and a capitalist economic system. With UNEP located in a Southern country and potentially inclined to favor the needs of the largest voting block inside the UN GA, the reality of the power politics at the time meant that UNEP could not afford to ignore the will, or the funding, provided by the United States. Consequently, a key pattern within international environmental affairs emerged. Through their numerical superiority within the UN GA, Southern states may dominate when determining the agenda for a conference. However, wealthy states, particularly the United States, may utilize their funding to control the actual outcomes of environmental affairs.

While realist theory certainly supports ideas such as state sovereignty and the importance of creating wealth within a county, this theory does not have full explanatory power for international environmental affairs. For example, the linkage between industrialization, wealth, and environmental damage suggests that states continuously pursue economic growth without considering environmental impacts. Thus, there is a contradiction between realist thought and the actual outcome of the Stockholm Action Plan and the Stockholm Declaration. States are cooperating, at least superficially, to delink industrial growth from adverse environmental impacts. Consequently, studies of international environmental diplomacy tend to utilize theoretical approaches that focus on cooperation.

3.2 UNEP Goes to Work

States tasked UNEP with monitoring humankind's impact on the environment by collecting quantitative data that measured the rate of ecological change and degradation. The first UNEP Governing Council meeting occurred in Geneva on June 12–22, 1973. UNEP immediately moved to establish a monitoring system to implement the Stockholm Action Plan. The North–South gap proved to be the dominant divide within environmental affairs [2]. This is further confirmed by the decisions made at the meeting that prioritized work on human settlements, land and water management, and assistance to Southern countries, including technology transfer, development aid, educational programs, and information sharing. Strong noted that funding for UNEP did not match the estimated 100-million-dollar price tag but that funding for the first year had been deposited [2].

UNEP turned quickly to acquiring data by which to measure the actual status of the global environment in the form of convening the first UN Intergovernmental Monitoring Meeting from February 11–20, 1974. This unenviable task of collecting and assessing data that reflects the global state of the environment occurred under the moniker Earthwatch. Earthwatch did not create an entirely new global monitoring system but was envisioned as a data hub maintained by UNEP and freely shared between countries. Thus, Earthwatch incorporated existing national and international monitoring systems while providing technical expertise, assistance in training, and monitoring equipment to developing countries. Additionally, Earthwatch sought to monitor the environment in those areas known as the commons, outside of the jurisdiction of the state system, such as oceans and space [35].

Major environmental monitoring systems incorporated into Earthwatch include the Global Environment Monitoring System (GEMS), the International Referral System for Sources of Environmental Information (Infoterra), and the International Register of Potentially Toxic Chemicals (IRPTC). GEMS constitutes the data collection system, while Infoterra focuses on exchanging environmental information between countries. The Stockholm Action Plan suggested the creation of the IRPTC database in Recommendation 74(e). The second UNEP Governing Council, held in Nairobi in March 1974, formally authorized UNEP to begin work on GEMS, Infoterra, and IRPTC.

Jensen et al. [19] subdivided Earthwatch's activities into four parts, including monitoring environmental status, conducting original research as needed, evaluating the quality of the environment, and exchanging information between countries. This system heralded a dramatic increase in cooperation for environmental affairs. Using the adage that "you cannot manage what you cannot monitor," a data management system for the environment carried with it the hopes and ambitions for much of UNEPs future work. Thus, it may be more understandable why the UNEP Governing Council authorized work on this system as a priority action item.

Gosovic [10] wrote that by 1992, GEMS became synonymous with Earthwatch while the other two systems, Infoterra and IRPTC, retained a lesser status. States linked GEMS priorities directly to the newly created UNEP agenda, including atmospheric pollution and climate change, potential contamination of food with chemicals, ocean pollution and its impact on marine ecosystems, factors needed to forecast natural disasters such as hurricanes, earthquakes, or tsunamis, or other adverse effects on human health. Further, as the possessor of global environmental data, UNEP should also be among the first to sound the alarm at any threats or crises involving environmental quality.

Infoterra, the second of three major systems within Earthwatch, fulfilled Stockholm Action Plan Recommendation 101 when it began full operations in January 1977. Infoterra operates as a Programme Activity Centre within UNEP. Affiliated networks designate national focal points that voluntarily share data. The data is sent to Infoterra in Nairobi for processing. This data may be requested via another national focal point within the partnership.

The Infoterra system design encourages states to develop their own monitoring system to participate fully. States recognized that some would not be able to provide the UN system with the reports necessary to determine the state of the global environment. Accordingly, UNEP provided developing countries with the necessary technical and financial assistance to create national systems to collect the data and write reports detailing the status of their domestic environment.

According to Villon [36], Infoterra stimulated both an improvement in environmental quality and better-informed decision-making. Further, access to adequate data allows national governments to make more informed choices about the restrictions needed to protect new pollutants from entering the environment. Given these benefits from utilizing Infoterra, other international organizations within the UN system also participate in the Infoterra system.

The final system within Earthwatch, IRPTC, began operations in 1979. Physically housed at the WHO in Geneva, the IRPTC collects and disseminates information about the impact of chemicals on both humans and the environment. This data includes potential chemical hazards, including national and regional regulations of potentially harmful chemicals. As part of this data system, scientists should also be able to identify missing information about chemicals and their impact on the environment in order to design new research projects that better inform humankind about the costs and benefits of the continued use of these chemicals. IRPTC also maintained a network of national and private institutions interested in using this data.

States recognized the enormity of the data needs surrounding environmental management and planning globally. In some ways, states charged UNEP with an impossible task. States themselves were not equipped to capture all of the data needed to effectively manage the relationship between industrialization processes, social organizations, technology, and ecological systems. While Infoterra began sharing environmental knowledge with states, it did not necessarily facilitate the research to fill in the knowledge gaps. This research typically occurs outside the state system, although wealthy states typically fund research projects. Consequently, knowledge of interest to environmental diplomats may be widely dispersed and reside in unique places both inside and outside the state system.

While Earthwatch focused more on the scientific data needs, in practice, the knowledge base needed to make environmental policy also encompasses the fields of economics, international law and policy, political science, and sociology. Data required to complete international environmental treaty negotiations successfully may consist of knowledge of the environmental damage, causes of environmental damage including production technologies, readily available equipment to eliminate or mitigate environmental damage, projected impacts of the changes to the economic system, more commonly referred to as the cost of compliance, and sufficient knowledge of political and social systems responses to these problems.

The establishment of GEMS, in particular, likely benefitted from connecting to existing UN environmental monitoring stations operated by other specialized agencies. Joyner [20] identified at least eight other specialized agencies that monitored one or more aspects of the global environment, including the FAO, the ILO, and UNESCO, who operated a monitoring network that utilized river stations to determine the pollution discharge into oceans. In light of this existing work, UNEP's mandate to serve as a catalytic organization may be understood in part as a desire to avoid duplicating efforts and avoid a dispute over primacy within the UN system [16].

The collection of scientific data is not limited to UN specialized agencies. National governments also sponsor collection activities separate from the Earth watch system for various reasons. For example, the United States government established the Mauna Loa Observatory in Hawaii in 1958 to measure carbon dioxide long before global diplomats decided to add climate change to the international agenda.

In addition to the national governments, public and private research institutions such as colleges and universities also disseminate new knowledge about science and technology. Gray [11] distinguishes state-controlled scientific knowledge and knowledge that circumvents state control. He points out that states may prefer preventing the free flow of scientific knowledge in circumstances where the state may be harmed, or its negotiating position undermined. Part of UNEP's success stems from UNEP's ability to assemble teams of scientists to work together in a technical sense while avoiding many political entanglements.

Additionally, non-state actors also produce and utilize knowledge. Almost from the beginning of UNEP, this organization formed partnerships with NGOs that allowed these groups insider access to the personnel carrying out the myriad of tasks that range from developing a high-level strategy to collecting data about the state of the environment. World renown anthropologist Margaret Mead addressed the Governing Council of UNEP at its meeting in 1977, acknowledged UNEP's request for greater NGO participation, and agreed that NGOs should be complementary to UNEP as a vital part of the international system; one capable of providing specialist knowledge as well as focusing public interest on environmental concerns [26].

In addition to operating the Earthwatch system, UNEP also published reports on the state of the global environment. Beginning in 1974, UNEP published an annual report detailing the State of the Environment. However, at the fourth Governing Council session in 1976, the Governing Council directed UNEP to narrow its focus to a select number of topics per year and to produce a comprehensive report every five years. UNEP established criteria for including an environmental issue such as its international importance, the emergency of a new problem (or of new science that would require the reevaluation of the situation), the urgency of the problem, and the lack of international attention to the issue at hand.

Limits to UNEP's ability to complete these tasks materialized immediately. UNEP operates at a considerable disadvantage when attempting to understand new chemicals and new technologies. The creators of these new products and processes have no mandate to share information on their environmental impacts. Given the significant increase in technology change, keeping the data systems entirely up to date proved to be an impossible task.

Further, the funding mechanism for the entire Earthwatch system depended upon additional financial commitments from the Northern countries. This funding did not consistently materialize. Palme [28] reported that states contributed sufficient financing for 1974 and 1975 for Earthwatch; however, this funding dried up by the fourth session of the Governing Council in 1976 [29].

In line with the Southern concerns about their ability to industrialize in the future by growing their domestic economies or through development assistance from the North, the Stockholm Action Plan also included a data request to determine whether environmental protectionism could threaten the Southern economy, possibly through a ban on Southern exports [21]. The South sought this data as part of a strategy to extract new financial resources from the North that would potentially use the do no harm principle to argue for compensation that environmental protectionism caused distinct harm to the Southern economy.

In and of itself, the collection of data may not be politically noteworthy. The underlying premise, however, of cooperation may well be considered essential. The collecting and assessing a common set of scientific data may be utilized to argue for a liberalist model when assessing a rationale for the UN. The liberal theory takes as its point of departure that rational man cooperates for the good of all [23]. This mandate applies to individuals and their social organizations such as states, corporations, NGOs, and international organizations. Liberal thinkers assume that social progress is probable, given that a rational man will cooperate to secure good for all. Further, cooperation when maintained over time, brings about a better life for humankind [30].

While realist theorists tend to focus almost exclusively on state action, liberalist theorists also include the interactions between organizations. Keohane and Nye [22] advance the theory of complex interdependence that relations between states occur on more levels than from one head of state to another head of state. Other groups and individuals both within formal diplomatic circles as well as outside of formal diplomatic negotiations also interact. This contact decreases the emphasis on states, including the usefulness of military power within international relations. That is not to say that power as a concept becomes less valuable. Instead, Keohane and Nye [22] argue that other forms of power, such as negotiation skills, become more critical.

Hurd [14] believes that liberalism emphasizes actors' choices when seeking to accomplish their interests. Liberalism, then, emphasizes the agreements states (or other actors) make when they choose to cooperate with others. States enter treaties when the benefits from participating in a treaty outweigh the costs. Liberalist theory emphasizes countries as the primary actor seeking to protect their interests. According to liberal theorists, international environmental diplomacy occurs because the transnational movement of environmental pollution encourages states to cooperate. Eliminating pollution that crosses national boundaries requires active, willing participation of independent states. International organizations such as UNEP need only manage the system to encourage these states to follow the same principles of a good neighbor and do no harm.

Theoretically, states participate in Earthwatch because the benefits of receiving information about the environment outweigh the low costs of sharing information with others. However, this may not be accurate. Fritz [9] pointed out that the initial criticism of GEMS limited its activities to specific sectoral actions and prevented it from becoming as strong of a monitoring system as initially envisioned. Gosovic [10] theorized that while states generally sought information about the environment, they did not prioritize controlling that information flow. These two facts essentially meant that utilizing Earthwatch to increase cooperation for the sake of environmental protection did not materialize as anticipated [9].

Despite Earthwatch failing to live up to expectations, knowledge nevertheless emerged as one of the power sources within international relations. As knowledge possession is not limited to states, various actor types could harness this to improve their standing within international relations. Knowledge as a source of power combined with the distribution of knowledge across actors highlighted inequities in the distribution of power. The conceptualization of knowledge as power unequally distributed across actors becomes a critical difference between the realist and liberalist approaches to analyzing international relations theory.

3.3 Catalyzing Cooperation

The primary task states asked UNEP to accomplish over time was to catalyze cooperation, that is, to encourage states to take actions to improve environmental quality by halting ecological withdrawals and ecological additions. States left undefined the mechanisms by which UNEP should achieve this outcome besides the mandates of the Stockholm Action Plan and the new norms contained within the Stockholm Declaration. These two documents brought states together despite a deep division on the importance of emphasizing environmental protection on behalf of the Northern countries versus the development needs of the Southern countries. The conceptualization of eco-development as a way to overcome Southern resistance to environmental quality served, at least temporarily, as a means of reconciliation that allowed for the creation of both the formal structure and the informal norms of the international environmental movement.

The entire Stockholm Action Plan and the Stockholm Declaration combine to support this viewpoint. States, in recognition of the terrible consequences of unchecked industrialization, began the process of cooperating to make forward progress on a significant threat to the global population. The UN GA also confirmed the status of the Stockholm Declaration by adopting Resolution 2994 (XXVII) on December 15, 1972. Sohn [31] reviewed the diplomatic statements made in support of the Stockholm Declaration and provided a commentary on all the principles within the Declaration. Sohn's analysis showed that states' receptivity to the Stockholm Declaration was decidedly mixed. The document contained various normative statements designed to appeal to both sides of the North–South gap. Perhaps unsurprisingly, the Stockholm Declaration included calls for state sovereignty and emphasized the need for states to manage environmental affairs in such a way as to minimize the impact on other surrounding states, one of the few issues on which countries on both sides of the North–South gap agreed. These ideas are reinforced in the UN GA Resolution 2995 (XXVII) and Resolution 2996 (XXVII) that emphasizes the "do no harm" principle of international common law, including the right of victims to ask for and receive compensation for the harm that occurred.

Another principle that states on both sides of the North–South gap agreed upon was the insistence on future economic growth. Countries in the South insisted upon strong language indicating their right to future development as a means of securing their economic and political independence. A careful reading of this Declaration reveals that the document places a stronger emphasis on the concerns of the developing South relating to economic development, social justice, and parity with the developed North than on emphasizing pollution control. Southern concerns about economic development, anti-colonialism, and social justice resonate throughout the document but are especially prominent in Principles 1, 8, 9, 10, 11, 12, 15, 20, and 23 [34]. In stark contrast, Principles 6 and 7 call directly for eliminating pollution, although many of the other principles recommend careful environmental planning and management.

Additionally, the Declaration emphasizes that all states are responsible for the quality of the environment in their own countries, but that states have different obligations based on the amount of development domestically [34]. This idea of "common but differentiated responsibilities" notes that underdeveloped states may not have the ability or even capability to protect the domestic environment. This concept acknowledged the sovereign equality of states while also acknowledging the difference in states' economic wealth, development status, and capacity to implement new programs in general. The articulation of this idea also included the request that developed states provide financing and technology transfers to assist developing countries in meeting their obligations stemming from implementing these new norms.

Other important environmental principles that appeared in the Declaration include the necessity of utilizing the best available science for the basis of international environmental treaties, encouraging environmental education of future generations, the need to incorporate the wise use of natural resources as well as the need for environmental planning and management, non-discriminatory population policies, and an end to nuclear weapons [34]. Of these remaining principles, states highlighted improving environmental planning and management at state and international levels.

Last, in the absence of new international law, UNEP provides guidance on creating new normative values and beliefs through publishing recent reports containing guidelines and suggestions for states. Engfeldt [7] pointed out that questioning traditional values in order to incorporate quality of life issues was appropriate in light of the experiences of the Southern countries. This question continued, and in 1980, the World Conservation Strategy contained the first reference in an official UN document to sustainable development, an alternative paradigm to the current capitalist economic system.

In addition to establishing new norms for environmental protectionism, both the Stockholm Action Plan and the Stockholm Declaration set out to deliberately expand the realm of actors working on this goal. While, on the one hand, states hold primacy within the international state system, the information demands, technological innovations, and social changes far exceed the ability of states to force change through their domestic systems. Thus, states left open the possibility of UNEP involving NGOs as part of its call to catalyze cooperation within international environmental diplomacy. Egelston [6] wrote that in the aftermath of UNEP's creation, the organization established relationships with NGOs willing to support eco-development norms. In doing so, NGOs gained direct access to decision-makers and shaped the discourse of international environmental politics.

In contrast to the realist and liberalist viewpoints, social constructivism believes that environmental problems have meaning as defined by how people perceive the world [12]. According to constructivists International relations revolve around social theories people hold about how the international state system functions. The international state system exists because people have ideas about how the international state system should work. People's ideas and beliefs about the global system drive the entire state system, how it operates, and what it could accomplish. Thus, ideas, values, beliefs, and assumptions matter within the constructivist theory.

Change occurs when the underlying ideas people hold about how the state system should enact change. Finnemore [8] argued that norms promoted by international organizations convince states to change behaviors. Norm entrepreneurs emerge that support the norm and work toward ensuring that others adopt the norm as a precursor to changing their behavior [32]. Barnett and Finnemore [3] further stated that international organizations possess a unique moral authority that positions them to accomplish this task. While the Stockholm Declaration defined norms for the international environmental agenda, these norms did not deviate from the traditional emphasis on state sovereignty or significantly impede the economic growth mandate. Instead, both the Northern and the Southern countries agreed on the necessity of maintaining state sovereignty. Additionally, the developing countries successfully ensured that the environmental norms embedded within the Stockholm Declaration could not be used to block the industrial growth of the developing countries.

Kuhn [24] asserted that new paradigms emerge from old ones. Thus, one of the contemporary debates within international environmental diplomacy involves whether the eco-development process should be considered a new paradigm that competes with the current capitalist paradigm. The answer to this question is a resounding no. There is nothing within this chapter to suggest that eco-development rose above the status of an exciting idea that might be useful in overcoming political resistance at a specific moment and place within international environmental diplomacy.

3.4 Forward March?

From the very beginning of UNEP, scholars such as Kwasniewski [25] questioned the anticipated effectiveness of UNEP. Scholarly opinion on UNEP's role, function, and effectiveness hardened immediately. An issue within academic circles is the standard by which UNEP should be measured. Effectiveness could be ascertained by determining whether UNEP has a strong possibility of improving environmental quality. Assessing this at the time of UNEP's creation is admittedly difficult. Engfeldt [7] believed that the global environment required a global rethink of the structure and organization of modern society with its emphasis on wealth accumulation and unequal distribution of the benefits of industrialization. He believed that

UNEP had the correct structure and flexibility to overcome what was the largest hurdle preventing further actions from improving environmental quality and state sovereignty [7].

As Von Moltke [37] pointed out, the vast majority of what would become the environmental agenda—ozone depletion, climate change, hazardous waste reduction, and biodiversity did not exist in any meaningful way during this time. Northern states articulated international environmental concerns as acid rain, halting whaling, and an emerging concern about ocean pollution. It is, therefore, less surprising that Southern countries successfully added their environmental concerns to this agenda: improvements in habitat, access to clean water and reliable energy, and the ever-present call for additional development aid.

Given the divergence of the issue areas assigned, UNEP's mission is admittedly unique. No other agency before or after was given the mission impossible task—to coordinate the work of older, more robust, better funded, and more sophisticated agencies from a location removed from the hallways of power. Scholars writing contemporary to the founding of UNEP thought that the creation of UNEP heralded a "modest" departure from current UN activities [5, 13, 20, 37].

These criticisms did not disappear over time. UNEP's critics argue that the program was hamstrung by its location, mandate, funding mechanism, and institutional weaknesses [10, 11]. Von Moltke [37] disagreed. He acknowledged that significant weaknesses indeed hamstrung UNEP; however, given that environmental concerns successfully trumped the development agenda in the 1990s, UNEP should be considered surprisingly successful given its inauspicious function within the international system.

Ivanova [16] agreed and stated that historical accounts that view UNEP as a maligned organization placed in Nairobi, Kenya, for the malevolent purposes of sidelining this new organization fail to account for more altruistic motives. She instead argued that the normative and catalytic functions demanded of UNEP plugged a gap in the institutional infrastructure of the UN without infringing upon other specialized agencies that already had specific mandates such as UNESCO, the WHO, or the WMO. UNEP functioned and continues to function as an anchor institution for the environment, meaning that the institution carries out an important role within the international system by collecting and analyzing data as well as by providing technical knowledge and political skill to broker agreements that appropriately solve the underlying problem [15, 17, 18]. Ivanova [18] concluded that UNEP successfully achieved much of the admittedly limited goals proscribed in its institutional mandate.

A third view espoused by Caldwell [5] ascertained that the need for success in controlling and reversing environmental damage was so great that states should not have given this task to a single organization alone. Nor, in practice, did the UN system leave environmental affairs solely to UNEP. Various specialized agencies within the UN system and other important intergovernmental organizations such as the World Trade Organization (WTO) contribute to protecting our global environment.

Like much of international environmental politics, no single vision for UNEP became a reality. Therefore, it is not surprising that scholars, practitioners, and students of this arena disagreed and continue to disagree on whether UNEP functions appropriately. Indeed, given that scholars disagree on the fundamental models that accurately represent international affairs, differences in opinion as to the successes or failures of the environmental conferences are inevitable.

References

1. Abbott KW, Snidal D (1998) Why states act through formal international organizations. J Conflict Resolut 42(1):3–32
2. Ambio News Briefs (1973) Ambio 2(4):125–128
3. Barnett M, Finnemore M (2012) Rules for the world. Cornell University Press, Ithaca
4. Belcher O et al (2020) Hidden carbon costs of the "everywhere war": logistics, geopolitical ecology, and the carbon boot-print of the US military. T I Brit Geogr 45(1):65–80
5. Caldwell LK (1972) Defense of earth in a divided world. J Environ Health 35(3):228–236
6. Egelston AE (2012) Sustainable development: a history. Springer, New York
7. Engfeldt L-G (1973) The United Nations and the human environment—some experiences. Int Organ 27(3):393–412
8. Finnemore M (1996) National interests in international society. Cornell University Press, Ithaca
9. Fritz J-S (1997) Earthwatch 25 years on: between science and international environmental governance. IIASA Interim Report. IIASA, Laxenburg, Austria: IR-97-059
10. Gosovic B (1992) The quest for world environmental cooperation: the case of the UN global environment monitoring system. Routledge, London
11. Gray M (1990) The United Nations Environment Programme: an assessment. Environ Law 20(2):291–320
12. Hannigan JA (1995) Environmental sociology: a social constructionist perspective. Psychology Press, London
13. Hardy M (1973) The United Nations Environment Program. Nat Resour J 13(2):235–255
14. Hurd I (2020) International organizations: politics, law, practice. Cambridge University Press, Cambridge
15. Ivanova M (2005) Can the anchor hold? rethinking the United Nations Environment Programme for the 21st century. Yale School of Forestry and Environmental Studies. https://elischolar.library.yale.edu/cgi/viewcontent.cgi?article=1026&context=fes-pubs. Accessed Jan 2 2022
16. Ivanova M (2007) Moving forward by looking back: learning from UNEP's history. In: Swart L, Perry E (eds) Global environmental governance: perspectives on the current debate. Center for UN Reform Education, New York, pp 26–47
17. Ivanova M (2009) UNEP as an anchor organization for the global environment. In: Biermann F et al (eds) International organizations in global environmental governance. Routledge, London, pp 165–187
18. Ivanova M (2021) The untold story of the world's leading environmental institution: UNEP at fifty. MIT Press, Cambridge
19. Jensen CE et al (1975) Earthwatch. Science 190(4213):432–438
20. Joyner CC (1974) Stockholm in retrospect: progress in the international law of environment. World Aff 136(4):347–363
21. Kennet W (1972) The Stockholm conference on the human environment. Int Aff 48(1):33-45
22. Keohane RO, Nye JS (1977) Power and interdependence: world politics in transition. Scott, Foresman and Company, Glenview
23. Keohane RO (1989) International institutions and state power: essays in international relations theory. Westview Press, New York
24. Kuhn T (1970) The structure of scientific revolutions, 2nd edn. University of Chicago Press, Chicago
25. Kwasniewski K (1974) Trouble with the UNEP. Inter Econ 9(5):130

26. Mead M (1977) Human settlements and the UN Environment Program: the potential of the non-governmental organizations (ngos). Ekistics 44(262):177–183
27. Morgenthau HJ (1965) Scientific man versus power politics. Phoenix Books, Chicago
28. Palme T (1974) Confrontations and practical action at UNEP meeting. Ambio 3(2):95–96
29. Palme T (1976) Financial difficulties and program review at UNEP meeting. Ambio 5(3):143–147
30. Pollard S (1971) The idea of progress: history and society. Penguin, Harmondsworth
31. Sohn LB (1973) Stockholm declaration on the human environment. Harvard Int Law J 14(3):423–515
32. Sunstein CR (1996) Social norms and social roles. Columbia Law Rev 96(4):903–968
33. UNCHE (1971) Report of the secretary-general to the third session of the preparatory committee to the UN Conference on the Human Environment. A/CONF.48/PC11
34. UNCHE (1972) Report on the United Nations Conference on the Human Environment. A/CONF.48/14/Rev.1
35. UNEP (1974). Approval of activities within the environment programme, in the light, inter alia, of their implications for the fund programme. UNEP/GC/24
36. Villon AF (1980) INFOTERRA: a global network for environmental information. Environ Int 4(1):63–68
37. Von Moltke K (1996) Why UNEP matters. In: Bergesen HO, Parmann G (eds) Green globe yearbook of international co-operation on environment and development 1996: an independent publication on environment and development from the Fridtjof Nansen Institute. Oxford University Press, Norway, pp 55–64
38. Wendt AE (1987) The agent-structure problem in international relations theory. Int Organ 41(3):335–370
39. Wygant CG (2004) The United Nations conference on the human environment: formation, significance and political challenges. Thesis, Texas Tech University

Chapter 4
Oceans, Seas, and Whales

Abstract Garrett (Hardin in Science 162:1243–1248, 1968) *Tragedy of the Commons* provides the backdrop for investigating a series of ocean themed international environmental treaties including the Convention for the Protection of the Mediterranean Sea Against Pollution as an exemplar of the Regional Seas Program, the United Nations Convention on the Law of the Seas and the International Whaling Commission's moratorium on whaling. From the Law of the Seas that defined the boundaries between national jurisdictions and the commons, protecting common pool resources through international treaty making stands out as the dominant theme of this chapter. Other examples include the Regional Seas Program, once referred to as the crown-jewel of UNEP that motivated states to participate more fully in the fledgling organization. In stark contrast, the International Whaling Commission struggled to enact a moratorium on whaling, despite the support of environmental non-governmental organizations and the public. This chapter also provides a brief synopsis of regime theory with an emphasis on regime formation to examine the differences between these three international treaties.

Keywords Tragedy of the commons · United Nations Convention on the Law of the Seas · Whaling · Regime theory · Regional Seas Program · Convention for the Protection of the Mediterranean Sea Against Pollution

From the beautiful beaches of France and Spain to the picturesque scenery of Greece and Italy, the Mediterranean Sea has been intricately involved in the history of world civilizations. The sea provided the backdrop for the ancient empires of Egypt, Greece, and Rome. At various times, the Mediterranean has been a daunting barrier between empires and a sea lane for travelers, a source of food for the poor dependent on the sea life for their very survival, and an exclusive playground for the global elite that could afford its expensive residences and picturesque views. Consequently, the Mediterranean Sea provides a variety of goods and services to the countries and peoples that share its shoreline. This complexity includes what type of ownership should apply to the body of water, the life below water, and the life on the shorelines.

This complex identity of what the Mediterranean represented not only for ancient realms but also for modern times not only existed in practice, it also existed in law. Ancient Roman thought also generated two legal principles of significant importance that apply to the Mediterranean Sea, *res nullius* and *res communis*. These two terms share the Latin term *Res*, which denoted an object that could be owned. *Res nullius* marked the absence of an owner, while *res communis* represented the idea that the property is used or owned by all. Beginning in the eighteenth century, states claimed sovereign control over a band of water off the coast of each country. The seas beyond that point became *res communis*, the property of all.

Who then becomes responsible for the environmental pollution in the Mediterranean Sea? States retain clear ownership of the pollution within their territorial waters closest to its shore. However, once pollution crosses this artificial boundary, the pollution joins the *res communis*. Common problems on the high seas include a variety of ecological ills ranging from the microplastics in the water to the nitrogen depletion that threatens aquatic life. This global mess combines with toxic agricultural runoff and oil spills from marine vessels to damage the aquatic ecosystem. The global elite gathered around the shores and on the waters do not want their playground damaged; the global poor, more directly dependent on the life below water, would not easily survive the collapse of these vibrant fisheries.

The Mediterranean Sea is not alone in its circumstances. It was, and still is a source of life, recreation, and transportation. Many other bodies of water, such as lakes, seas, and oceans, also are threatened by environmental pollution. Countries allow the problems that caused the environmental damage to the Mediterranean Sea to occur in waters worldwide. For example, scientists have been monitoring the Great Pacific Garbage Patch, a 1.6 million km area of the North Pacific Ocean between Hawaii and California filled with plastic debris that entered the ocean from rivers [23]. The plastic pieces travel through the water on gyres into the heart of the seas, where they eventually break down into microplastics. The microplastics block sunlight from entering the water, negatively impacting the growth rates of algae and plankton at the bottom of the food chain. This change in the food supply could carry up the food chain in the marine environment and cause devastating impacts on ocean life and other elements of the marine environment.

Ocean health, then, is of vital importance for life on the planet. Water covers 72% of the Earth's surface. Oceans form the marine environment and help maintain the delicate balance that allows life on land to thrive by producing oxygen and absorbing carbon dioxide. Oceans also assist in regulating our temperature and climate by moving heat between the equator and the poles. Additionally, oceans play a significant role in the global economy. Oceans provide food through fishing and the harvesting of other wild marine-based foods, serve as a repository of oil and gas and other seabed mining, and demonstrate a significant potential for renewable energy production, including offshore wind energy. Oceans also can provide freshwater when treated with a desalination plant that removes the salt from the water. The ocean environment serves as a popular destination for tourism as people flock to beaches for relaxation or participate in diving to explore life under the water.

In recognition of the many services provided by the oceans, economists, scholars, and activists began work on conceptualizing these relationships as part of the Blue Economy conceptualization that originated in the UN Conference on Sustainable Development in 2012. This conceptualization integrated economic development and social justice with a focus on the role of water [39]. The World Bank and the United Nations Department of Economic and Social Affairs (UN DESA) [55] described the Blue Economy as consisting of economic activities and the governmental decisions that ensure sustainability. Thus, the Blue Economy includes oceans and coastal communities that utilize ocean resources. Consequently, humankind should manage these resources that generate economic wealth in a manner that produces social justice for all.

Concerns about the health of the marine environment were already present at the beginning of the international environmental movement. Engfeldt [5] reported that states expressed considerable concern over marine pollution at UNCHE. Thus, Principle 7 of the Stockholm Declaration for the Environment and Recommendation 47 of the Stockholm Action Plan explicitly called for countries, fisheries agencies, and international organizations to prepare for the UNCLOS III to protect the marine environment. Concerns about the seas, fishing practices, or coastal areas appeared throughout the Stockholm Action Plan in Recommendations 7, 33, 47, 49, 55, 86, 89, and 90 [47].

This chapter reviews the actions taken at the international level to support the protection of our oceans and other global commons. The chapter opens by presenting a theoretical viewpoint of the oceans as a shared space beyond the jurisdictional claims of nation-states. This status triggers unique problems for the oceans and life within them, as no one group can control or protect the quality of the marine environment on the high seas. The second section reviews the events leading up to the moratorium on whaling under the auspices of the IWC. While unknown at the time, the actions of a select few states threatened large whales. Multiple whale species, including the right, gray, humpback, and blue whales, were hunted almost to extinction. The third section reviews treaties that deal with our ocean environment, beginning with the UNCLOS III that established the extent of national jurisdiction near shorelines and thus defined the extent of the high seas. The fourth section looks at the UN RSP that began with the Mediterranean Sea. The agreement negotiated from this process, the Convention for the Procedure of the Mediterranean Sea Against Pollution, served as an early prototype of success for regional agreements focused on preserving our marine environment while allowing reasonable economic usage. The fifth section illuminates the conceptualization of regimes as a mechanism for analyzing environmental treaties. This concept is essential in international relations approaches to modeling this field as regime theory became one of the dominant strands of academic analysis.

4.1 Tragedy of the Commons

Beginning with the UNCHE meeting in 1972, people realized that industrial activities generated significant environmental pollution that would significantly damage the biosphere if left unchecked. Writing contemporary to the origins of the Stockholm Conference, Hardin [13] articulated the Tragedy of the Commons. This metaphorical viewpoint established that individual action depletes open access resources without social controls or privatization. The potential destruction of the commons illustrates a social dilemma between preserving public land for all versus the economic rationale to consume resources by an individual.

Hardin begins by pointing out that each sheepherder would profit from grazing sheep in a public space as the herdsman would be relieved of providing an equivalent space from his own funds [13]. Thus, every herder in the area would place their sheep in the public space, leading to overexploitation and collapse of the property. In this case, an individual acting upon his own rational economic interests ruins the land held by a group of people without their input or consent. This hypothetical case exemplifies environmental affairs.

Scholars today recognize four global commons: the atmosphere, Antarctica, the high seas, and outer space. Common pool resources represent situations where a finite natural resource may be consumed because the cost of excluding a user is high [29]. These resources suffer the tragedy of the commons. Because no one owns these environments, individuals and organizations overconsume these resources. Whaling on the high seas meets the definition of a common pool resource since no country owns the mammal. Fleets of vessels were free to pursue whales as patrolling an ocean to look for whaling vessels could only be undertaken with significant cost and difficulty.

Solutions to the tragedy of the commons focus on how to constrain the rational self-interest of the polluter. Hardin [13] warned against relying on individual conscience as a means of protecting the commons, preferring top-down regulations instead to manage the resource. In other words, he prefered that the commons remain untouched backed by the state's authority to enforce compliance with laws and regulations managing access and resource usage. He also acknowledged that privatization ends the commons dilemma, albeit by removing the absence of ownership. The ability to exclude others from using the once-common resource incentivizes the new owner to protect the property.

American political economist Elinor Ostrom proposed a different solution to the tragedy of the commons. Ostrom [28] recognized that users of the commons frequently devise rules that balance the interests of all the users so that the public space does not become depleted. Each user then engages in self-policing, at times with substantial cost involved. In this respect, Ostrom solved the tragedy of the commons. In recognition of her work, she won the Nobel Prize in Economic Sciences in 2009, becoming the first woman to do so for this field.

Within the realm of environmental diplomacy, states embedded within the environment can potentially end the overconsumption of the commons through the creation of international environmental treaties that define a set of rules that spread the costs and the benefits of preserving the commons across all states. A robust enforcement system becomes necessary to ensure that all states adhere to the agreement.

4.2 Whaling

Embedded within the Stockholm Action Plan, Recommendation 33 suggested that governments agree to a 10-year moratorium on whaling [47]. By the early 1970s, scientists and the general public became concerned about the noticeable lack of whales in the oceans. The right whale earned its name as whalers found it easy to hunt, kill, and retrieve this whale from the water. Its large size and buoyancy meant that it floated on the surface of the water after its death. In addition to the right whale, other large species faced extinction. In 1970 the United States classified multiple whale species as endangered under the Endangered Species Conservation Act of 1969, a predecessor law to the Endangered Species Act. Six of the whale species listed as endangered included the blue, humpback, right, gray, sei, and fin. This law meant that products made from whales could not be brought into the United States without a permit.

Early attempts to regulate whaling at the international level began before World War II in an attempt to protect the right whale. The efforts were short-lived due to the necessity of acquiring food after the war. Whales not only provided food but also appeared in other products. Whale blubber was an ingredient for whale oil, suitable in soaps, waxes, and candles. The IWC came into existence in 1946, when fifteen members interested in managing whales to ensure the future of commercial whaling agreed to collaborate in order to set quotas to preserve the whale stock.

States used a mechanism to estimate the appropriate whaling quotas based upon a blue whale equivalency method. Instead of assigning a quota per country, the IWC determined a limit for all countries that whaled for the year. Thus, countries that processed the whales the fastest consumed the largest portion of the quota for that year. In this sense, the IWC incentivized quickly catching as many whales as possible. Unsurprisingly, this mechanism allowed whaling countries to meet estimated market demand while simultaneously causing the number of whales to continue to decline [9, 27]. By the late 1950s, the failures of the quota system became apparent as states proved unwilling to reign in their whaling industries resulting in a failure to determine the quotas in the 1959/1960 pelagic season. Additionally, this quota system utilized a weak enforcement system. The IWC requested observers to join the whaling fleet, but observers' powers were limited to reporting whether they believed the whaling data accurately reflected the actual catch [17, 54].

By 1972, states switched the mechanism used to establish the whaling quotas from an economic output-based count based on blue whale oil equivalents to a maximum sustainable yield mechanism. The maximum sustainable yield methodology relied on scientific estimates of the overall size of the population to estimate how many whales could be killed without endangering the survival of the species. Errors in the estimated maximum sustainable yield calculations could (and did) move the whales closer to extinction.

Public concern over whaling stemmed from their status as a beloved cultural icon, along with the growing realization that the IWC did not protect the whales as intended. NGOs reflected this public concern as they played a central and significant role in the efforts to ban whaling. Several nascent environmental NGOs turned their attention toward this issue, such as the World Wildlife Fund (WWF), Greenpeace, and the International Union for Conservation of Nature (IUCN). In part because of this campaign, these three NGOs established themselves as global leaders in environmental protection [35, 38, 40].

Sir Julian Huxley, a British evolutionary biologist and first director-general of UNESCO, became a founding member of the WWF along with Sir Peter Scott. Huxley also formed the IUCN, a global NGO consisting of member organizations that support conservation, including states, businesses, and other environmental NGOs. IUCN's Red List, begun in 1965, served and continues to serve as a data repository for species, including identifying critically endangered and extinct species such as large whales.

Four years after Greenpeace organized in 1971 to protest US nuclear testing, this NGO interfered with a Soviet Union whaling hunt in 1975. This incident began the international movement to enact a moratorium on whaling. Greenpeace, utilizing a yacht to sail to the general vicinity of the Soviet Union fleet, placed individuals into smaller inflatables that attempted to force the vessels away from the whales, and in the process, captured the incident in video footage [40, 58]. The publicity stunt that launched the "Save the Whales" campaign, when combined with scientific expertise from former Vancouver Aquarium researcher Dr. Paul Spong, hurdled Greenpeace into international prominence.

The "Save the Whales" campaign consisted of a multifaceted attempt to enact a moratorium on killing endangered whales. In order to enact the ban, NGOs worked on shifting public opinion in the United States and around the globe about the desirability of consuming whale meat and whale oil. The campaign included public education efforts, passing domestic legislation that either banned specific hunting methods or otherwise increased the costs of whaling. One amusing example of the campaign included the adaptation of a board game by the same name. Intriguingly, the game not only hit the stores but also became the focus of academic reviews about whether this type of activity could be included in a classroom [53].

Greenpeace did not act alone in this movement. WWF funded scientific research at Patagonia, Argentina, by Roger and Katy Payne that led to recordings of whale sounds [35]. These sounds immediately resonated with a fascinated public. They added to the growing sense that whales were not merely dumb fish but rather sophisticated animals integral to the ocean environment. Additionally, IUCN, with its political insider connections, hosted a joint workshop with UNEP on whale sanctuaries, entitled a Workshop on Cetacean Sanctuaries.

Perhaps the most infamous incident of an NGO directly engaging whaling ships involved the Sea Shepherd skippered by Paul Watson of Canada. Formally organized in 1977, the Sea Shepherd Conservation Society sought direct conflict with whaling ships that sought to circumvent existing IWC whaling quotas. One of the original Greenpeace members, Watson, took a more militant approach to end whaling than the already confrontational Greenpeace. In 1979, the Sea Shepherd rammed the whaling ship Sierra, a so-called "pirate" whaling ship off the coast of Portugal. The Sierra limped back to port, Watson and the other two crew members of the Sea Shepherd were charged under Portuguese law but fled the country. The Portuguese government seized the Sea Shepherd, probably with the intent of sending the boat to salvage to recompense Sierra's owners [36].

Seychelles, an island-state off the coast of East Africa, set an example by assembling a coalition to create the first whale sanctuary in the Indian Ocean [7]. This effort succeeded, and in 1979 the IWC voted to create the Indian Ocean Whale Sanctuary. Disappointingly, the sanctuary's boundaries did not extend into Antarctica, ending instead at the 55 degrees South parallel. This decision excluded the minke whale grounds in Antarctica, leaving this species vulnerable to whaling expeditions [15].

The dramatic shift in global opinion encouraged the United States to take a strong stance in support of banning commercial whaling. Domestic public opinion did not favor the continuation of whaling. Not only did the Greenpeace campaign educate the public on the need to conserve the whales, but it also tapped into the patriotism present during this era. The Greenpeace campaign that targeted the Soviet Union in the role of the antagonist suited Cold War sensibilities by utilizing the Cold War rhetoric that saw the United States as the defender of the noble, good, and just against what would later be labeled the "evil empire" of the Soviet Union by President Ronald Reagan in 1983 [38].

This "evil empire" cliché, unfortunately, rang accurately. Today, scientific and anecdotal evidence suggested that the Soviet Union's whaling fleets decimated whale populations, including the blue and humpback whales [59]. Sadly, these majestic animals did not die to serve as food but instead were mercilessly hunted down as the five-year Soviet plan called for their extinction [1]. At the time, the Soviet Union went to great lengths to hide the extent of their whaling and falsified reports to international agencies [17]. While other states suspected the Soviet Union was underreporting their whale catch at the time, other states did not challenge the Soviet Union for fear of undermining the IWC [54].

To counter the political and economic clout of the whaling industry, environmental groups also convinced new states to join the IWC for the purposes of enacting the badly needed moratorium on whaling. This effort succeeded as 20 new states joined the IWC between 1979 and 1983, transforming this international organization from focusing on the economic interests of the whaling fleets to conservation interests [7]. The increase in membership also restored public confidence in this organization.

The constant pressure on the IWC along with the further decline in whales convinced member states to vote for the moratorium on whaling [25, 37]. On July 23, 1982, the ban on commercial whaling passed the IWC to be placed into effect for the 1985/1986 season after securing a two-thirds majority vote of the member countries. This vote did not, however, end whale hunting.

The formal rules of the IWC recognize state sovereignty and include a mechanism for states to avoid regulations by filing an objection within 90 days of the IWC finalizing a rule. States with a strong cultural attachment to whaling filed an objection within 90 days, including the Soviet Union, Norway, Japan, and Peru [52]. These states took advantage of an IWC rules that require whales killed during scientific research not to go to waste. Whales could be hunted for research purposes, with the whale meat entering the supply chain for food or other commercial products.

Additionally, states could avoid the moratorium by commissioning the "pirate" ships. These ships, flagged out of a country that is not a member state of the IWC, would not be subject to a ban. The fleet could then catch an unlimited number of whales and sell the resulting products for a profit on behalf of their owners.

NGOs did not end their public pressure after the ban passed. Sakaguchi [37] investigated whether the actions of the NGOs hardened resistance to implementing the moratorium on whaling in Norway, Japan, and Iceland. NGOs, especially the more militant minded groups like the Sea Shepherd Conservation Society, continued to confront whaling fleets directly. Other more moderate NGOs publicized the plight of the whales and encouraged individuals to not only boycott whale products but also to avoid other commercial catches such as fish. Sakaguchi [37] concluded that NGOs' decision to pressure these three states through direct action and pressure tactics likely entrenched their whaling activities.

Today, several species of whales show promising signs of recovery from their endangered species status, while other species remain endangered due to scientific whaling. The IUCN Red List [18] noted that the humpback whale and the gray whale rate the least concern while the blue whale remains endangered, and the right whale persists in the critically endangered category. The minke whale, now the focus of Japan and Norway's whaling fleet, appears in the least concerned category.

4.3   Law of the Seas

Like the whaling regime, the UNCLOS III negotiating history began in the pre-Stockholm era. States' recognition of the need to establish a set of legal principles and treaty language covering the oceans dates to the early seventeenth century when Hugo Grotius, a Dutch philosopher, published *Mare Liberum* in 1609, a book arguing that the seas should be free and open to all.

The idea of mare liberum did not prevent the states from attempting to regulate the use of the seas. As part of the Pax Britannica, Great Britain allowed countries to claim a three-mile radius off their coasts as part of the national jurisdiction. However, the United States, as the remaining naval power after World War II, chose not to limit themselves or others to this narrow band of water. Instead, the United States, in 1945, laid claim to its continental shelf not only for fishing but also for seabed exploitation, including oil and gas drilling. Unsurprisingly, other countries followed suit.

Countries sought to contain these problems through the Geneva Convention on the Law of the Sea in 1958, sometimes referred to as the United Nations Convention on the Law of the Sea (UNCLOS). This treaty contained four separate conventions that combined to form a regime governing the high seas, including distinguishing territorial waters from the high seas and convening rights of access and removal of living resources such as fish and whales and natural resources such as tin and diamonds.

The Convention on the Territorial Sea and the Contiguous Zone attempted to settle the boundary between the high seas and territorial waters. This Convention also guaranteed the right of innocent passage of vessels passing through straits. The Convention on the High Seas codified customary international law; that is, the Convention put into writing the customary practices that guaranteed freedom of the seas. The treaty also confirmed the practice of flagging ships, respecting the freedom and independence of these ships, and allowing ships access to ports as requested. The Convention on Fishing and Conservation of the Living Resources of the High Seas attempted to put an environmental management system in place that allowed for the maximum amount of food production and strong safeguards for marine life conservation. The Convention on the Continental Shelf establishes states' sovereignty over the continental shelf, including the ability to remove and otherwise manage the natural resources in the seabed. This treaty acknowledged the expansion of the states' territories and, therefore, shrunk the high seas.

The Second United Nations Convention on the Law of the Sea (UNCLOS II) occurred from March 17–April 26, 1960. States met in an attempt to settle the two questions unresolved from the Geneva Convention in 1958; that is, the issue of the extent of the territorial jurisdiction and the resolution of allocating fishing rights. Conversations on convening UNCLOS III began when Malta's Ambassador Pardo, in a speech to the UN GA, expressed concern that the competition between states for resources such as tin, diamonds, oil, and gas, would lead to a steady depletion of the seabed [51]. As enacted by the United States and followed by others, the expansion of what would become Exclusive Economic Zones (EEZ) concerned Pardo. He encouraged the UN to create a new regime that governed the seabed outside national jurisdictions.

This speech heralded new negotiations that would eventually result in UNCLOS III. Formally, the UN established the Sea-Bed Committee in 1970 with 90 members as the initial negotiating committee [42]. This group laid the groundwork for negotiations to proceed, including investigating the issues at hand, assembling states' positions, and identifying the areas of both agreement and dissent.

By the time of the UNCHE meeting in Stockholm in 1972, three environmental concerns emerged stemming from the implementation of the Geneva Convention. First, states compete against each other to collect common pool resources such as fish and whales. Ambassador Tommy Koh of Singapore and President of the UNCLOS III noted that developing countries, in particular, were concerned about the common property resources becoming depleted as this is an inexpensive and reliable food source for many countries and peoples [19]. This competition may lead to the tragedy of the commons where these species become threatened with extinction.

Second, a variety of sources created contamination that accumulated in the seas [5]. This contamination can enter the waters through multiple pathways. Ships and marine vessels dump oil and sewage overboard as part of their operations. Also, chemical and biological wastes may also enter the oceans from shores, such as through a river or other land-based sources.

Third, states increasingly mine the seabed for various raw materials [5]. In other words, if a state owns the seabed, the state may either develop the area as part of its activities or assign the rights to a third party who could also exploit the site. Thus, the seabed, previously off-limits to offshore activities because they could not be claimed by any one group could now be divided among companies for the purposes of extracting raw materials.

Additionally, competition to claim these valuable resources could create conflict between states. Countries with stronger navies would have an advantage over other countries that did not possess naval power, exacerbating power differentials between countries. Brown and Fabian [4] added two additional issues needing resolution, such as revenue sharing stemming from the commercial development of natural resources contained within the seabed and a dispute resolution mechanism for complaints between states.

In addition to these substantive issues, Koh [20] reported that changes in the actors within international diplomacy also impacted these negotiations. The rise in the developing states wanted the opportunity to craft new international law in the hopes of acquiring new resources for their use. These states that were new to the UN used their new membership to revisit issues.

The first organizational meeting occurred in New York, at the UN Headquarters, from December 3–15, 1973. States developed the rules and procedures for the negotiating session, including the decision to work by consensus and to produce one package treaty rather than individual texts. Writing alone or in tandem, Stevenson and Oxman provided an overview of the negotiations for each meeting. They identified key issues such as defining the waters that should be within coastal state's jurisdiction, creating a new legal authority to oversee economic activities involving the seabed and crafting a dispute mechanism to resolve problems [30–34, 41, 42, 43].

Early in the process, diplomats hoped to complete negotiations quickly. Stevenson and Oxman [42], writing in their personal capacity while serving as members of the US delegation, expressed the opinion that UNCLOS III should conclude by the year's end. However, the negotiations lasted well over a decade, from the beginning of the UN Sea-Bed Committee in 1970 to opening UNCLOS III for signature in 1982. Fawcett [6] expressed that the length of the negotiations was unsurprising given the number of states involved, the complexity and importance of the issues, and the fact that international diplomacy historically moves slowly.

States completed negotiations on UNCLOS III in Montego Bay, Jamaica, on December 10, 1982. Notably, UNCLOS III establishes a legal regime for the high seas that contains provisions for environmental protection or made decisions that impact environmental quality. Diplomats expanded on the common heritage of humankind principle within the treaty. Additionally, states addressed the environmental concerns to varying degrees of success. The treaty also created a framework for equitable access to common pool resources and a new international agency to oversee the exploration and exploitation of the seabed. States also agreed to a dispute settlement mechanism for conflicts related to this new treaty.

UNCLOS III utilized the common heritage on humankind principle throughout the treaty. Frakes [8] articulated five elements within the common heritage of humankind principle. First, the absence of ownership applies. Second, the commons should be managed independently of all states through an international organization. Third, the commons should be treated as a global public good with any benefits from economic activities distributed equally across all parties. Fourth, no military activities should be allowed. Fifth, the area should be preserved for future generations to avoid the tragedy of the commons.

UNCLOS III settled the question regarding the boundaries between the high seas and territorial waters. The treaty allowed coastal states to claim up to 12-miles as territorial waters. At the end of the 12-miles, states could claim an EEZ over which it exercised control over all living materials and natural resources for an additional 200 miles. Stevenson and Oxman [41] report that extending the state's territorial waters appeared non-controversial because it also provided better management of the common pool resources. Instead, states concerns centered around access rights to the territorial waters for passage and the allocation process for catch quotas.

States overseeing their EEZ acquired the responsibility of managing common pool resources by implementing an allowable catch quota system for living marine resources while conserving the common pool resource for the future. Coastal states overseeing the EEZ could utilize the first claim to the catch quota, but they should prioritize landlocked states, followed by developing countries when determining the distribution of the quotas to fishing fleets. Additionally, coastal states also retained the right to authorize scientific research expeditions in their EEZs and territorial waters.

Koh [19] reported that both superpowers recognized that changing the width of the territorial waters from three miles to twelve miles potentially negatively impacted international navigation. Thus, UNCLOS III contained a new principle, transit passage, to reconcile these differences. States expressed concern about the conditions under which navies, particularly submarines, would utilize the straits. These straits, a narrow passage of water between two larger bodies of water, no longer formed part of the high seas but instead became part of the domestic waters of a county. Transit passage allowed submarines to move through the straight without coming to the surface and showing their colors, a necessary condition to claim innocent passage status. Additionally, transit passage allowed overflights as well.

Member states attempted to limit environmental pollution by agreeing to a single set of vessel pollution standards through the International Maritime Organization. This regulatory certainty benefitted the ship's owners and operators by confirming they would only need to comply with one set of rules. Additionally, flag states agreed to a uniform set of legal obligations. However, this system may prove easy to circumvent as these regulations create a race to the bottom regarding environmental, safety, and other regulations. especially when vessels may choose to operate under a flag of convenience that ignores internationally determined standards such as the International Convention for the Prevention of Pollution from Ships (MARPOL).

More controversially, states agreed to create the International Seabed Authority to exercise control of seabed mining activities within the high seas. Headquartered in Kingston, Jamaica, the International Seabed Authority ensures environmental protection for the 54% of the oceans considered part of the high seas. UNCLOS III established a legal regime where states and private companies seeking to explore the seabed for commercial activities must seek permission from the International Seabed Authority to do so. Seabed mining exploration occurs under an exploration contract that includes monetary transfers, either as payments or royalties, to the International Seabed Authority.

UNCLOS deviated from the pattern of weak enforcement measures frequently associated with international treaties in that states agreed to submit disputes to a third party for adjudication. United States law professor and member of the United States delegation Louis Sohn is typically credited for his work in this area [26]. If the disputing parties can agree on where to submit the dispute, then this mechanism supersedes the convention. Additionally, disputes may be adjudicated through various mechanisms, including: the ICJ, the International Tribunal for the Law of the Sea established by UNCLOS III for this purpose, ad hoc arbitration under Annex VII, or a special arbitral tribunal established by Annex VIII. Ad hoc arbitration under Annex VII serves as the primary dispute mechanism. Special arbitral tribunals only occur when the dispute arises because of military activity, an affair before the UN Security Council, or a maritime boundary dispute [48].

The above conversation focused, by necessity, on states as the primary actor. Particularly noteworthy in its absence, very few scholarly works mention the presence or activities of NGOs during the UNCLOS III negotiations. However, Levering and Levering [24] believed that NGOs impacted the negotiations. Of importance, however, is that Levering attended the conference as an accredited NGO and served as a leader in the Neptune Group, a conglomeration of NGOs working to ensure the success of the conference negotiations. Levering and Levering noted that NGO attendance, but not interest, in the negotiations was low due to the credentialing requirement that an NGO work internationally. More likely, however, is the idea that NGOs attempted, and may have succeeded in, influencing the negotiators at the conference but established scholarly researchers did not view this activity as worthy of scholarship contemporary to the events taking place.

After the UNCLOS III treaty opened for signature, support for the document did not quickly materialize. As is typical for international law, countries must first sign, then ratify a treaty. Countries that sign treaties agree to seek full ratification at an undisclosed future date and refrain from taking an action that conflicts with the rationale and contents of the treaty. Thus, signing a treaty can be considered a symbolic act of solidarity that does not create any legal obligations. Ratification occurred on November 16, 1994, when the sixtieth state submitted its ratification to the UN. As of 2021, 168 parties have ratified this treaty.

The United States, led by President Ronald Reagan, refused to sign the treaty. Stevenson and Oxman [44] noted that while the United States objected to Part XI that contained the provisions regarding seabed mining, among other topics, the United States complies with the remainder of the text. The United States, in 1994, did sign UNCLOS III under President Bill Clinton.

Despite the absence of the United States ratification, UNCLOS III's status within international law changed from a new treaty to customary law. Customary international law serves as one of the sources of international law. This form of international law arises from the customs and standards based upon the behaviors of states. In essence, customary law occurs because different states adhere to the same patterns of behavior.

4.4 UNEP's Regional Seas

From the beginning of the negotiations under the auspices of the Seabed Committee that began work on UNCLOS III, states realized a complementary system of bilateral and multilateral regional treaties would be necessary to adequately manage fishing populations and other relevant regional concerns [41]. UNEP's Regional Seas Program (RSP), sometimes referred to as the crown jewels of UNEP, provides this harmonization [16]. The RSP focuses its efforts on enclosed or semi-enclosed bodies of water that have a significant number of coastal countries that contribute to the quality of the water within the sea.

While the previous section reviewed threats to the sea from activities on top or underneath the water, pollution also enters the seas from land and waterways. This contamination does not occur because of the "commons" nature of the oceans themselves. Instead, this pollution occurs because sovereign states allow the damaging action to continue. Recognizing the increasing impact of this pollution on the Mediterranean Sea, states discussed the issue in the FAO's General Fisheries Council for the Mediterranean in early 1974 [22].

This conversation on limiting pollution to better conserve fish stocks for current and future consumption continued at the UNEP's Governing Council second session [49]. Spain, in particular, expressed an interest in hosting a meeting of states to discuss the topic further. Gosovic [10] reported that UNEP believed states would be interested in this arena as the pollution either occurred in their territorial waters or originated from pollution sources onshore.

The Mediterranean Sea is an enclosed body of water with an opening to the Atlantic due to the strait at Gibraltar at the western end and an entrance to the Marmara Sea through the Dardanelles in Turkey. Coastal countries occupy part of the continents of Europe on the northern coast, Africa on the southern coast, and Asia on the eastern shore. Countries participating in the Mediterranean RSP historically clashed culturally and militarily as the group contained members involved in mutual hostility, including Greece and Turkey, and the Arab–Israeli conflict. UNEP did, however, avoid adding more conflict by declining to invite the Palestinian Liberation Organization [11].

As with many environmental programs, UNEP followed a three-pronged approach of environmental assessment, environmental management, and supplemental activities such as environmental education, knowledge sharing, and technology transfer. Bliss-Guest and Keckes [2] pointed out that the RSP focused on four components. The first component specified management priorities, including control of pollution entering the water. The second component assessed the current quality of the environment, while the third component emphasized coordination of the protection, development, and management of marine and coastal resources. The fourth component supported states' engagement within the RSP by providing resources for the participation of interested countries.

UNEP allocated resources to monitor pollution sites close to shores that had the ability to detect specific point sources that may contribute to water pollution in the area. This pollution comes from multiple sources, the most significant of which include land-based pollution, with the remainder originating from vessels sailing the oceans. Helmer [14] noted that land-based pollution that entered the Mediterranean originated from industrial discharges, sewage, agricultural runoff that contained pesticides, and oil from marine vessels. Oil spills from marine vessels dominated the news cycle [45].

States accepting Spain's offer to host a conference in Barcelona met from January 28–February 4, 1975. Formally known as the Intergovernmental Meeting on the Protection of the Mediterranean, but more commonly referred to as the Barcelona Convention, states met to create the Mediterranean Action Plan (MAP). MAP consisted of four parts: a comprehensive plan for the development and management of resources, a coordinated research program, legal agreements, and institutional and financial arrangements [56].

Thatcher [45] stressed that the structure of the 1975 Barcelona Convention included both lawyers and scientists working under the direction of states, meeting in a plenary. UNEP [50] noted that action plans depend on scientific assessment to identify environmental problems for political processes to solve. MAP called for a network of scientific research institutions and a process by which interested parties could work together to develop the scientific knowledge and experience to measure and monitor pollution entering the Mediterranean. The Coordinated Mediterranean Pollution Monitoring and Research Program (MED POL) conducted the original environmental baseline assessments as a pilot program. This research program involved scientists from governments and other international organizations, notably WHO, FAO, Intergovernmental Oceanographic Commission, UNESCO, WMO, International Atomic Energy Agency (IAEA), and UNDP.

Thus, UNEP launched the MED X program in 1976 under the MED POL to determine the total amount of pollution load entering the Mediterranean, regardless of the country of origin. The MED X program assessed industrial wastes, sewer discharges, agricultural runoff, and river discharges. Scientists performing the emissions inventory considered radioactive releases from nuclear plants but dismissed them due to the low quantity of sources and emissions [14].

Haas [11] reported that UNEP funding not only added to the scientific knowledge about the area by providing funding to research that may not have otherwise received funding. It also cemented political alliances in support of the broader RSP. Haas' [12] research into the Med Plan led him to create the concept of epistemic communities, a group of like-minded scientists that framed environmental problems in similar terms, even though the scientists resided in different countries with different political interests. Epistemic communities held shared ideas about the causes of environmental pollution and the solutions to environmental pollution. Consequently, scientists advised their respective governments about similar problems and recommended similar solutions. Due to their prestigious place within states as technical advisors, these similarities assisted in reducing the differences in negotiating stances that allowed commonalities to become more readily apparent [12].

This pattern placed significant importance on the environmental assessment phase of international environmental cooperation. Scientists, rather than diplomats, would be the first to define the environmental problems impacting marine health. Thus, MED POL became deeply embedded into the actual operations of the entire Mediterranean regime. In addition to carrying out scientific studies regarding the health of the Mediterranean Sea, MED POL also made policy recommendations to states for further consideration.

The Conference of Plenipotentiaries of the Coastal States of the Mediterranean Region for the Protection of the Mediterranean Sea, established a simple goal for its signatories, keep pollution out of the Mediterranean Sea. States articulated this goal more fully in the three agreements from this meeting by accepting a framework convention, the Convention for the Protection of the Mediterranean Sea Against Pollution, better known as the Barcelona Convention.

Additionally, states negotiated two additional protocols. The Protocol for the Prevention of Pollution of the Mediterranean Sea by Dumping from Ships and Aircraft creates a blacklist of chemicals that may no longer be disposed of in the Mediterranean Sea and a grey list of chemicals that require a special permit to be dumped in the Mediterranean Sea. The Protocol Concerning Cooperation in Combating Pollution of the Mediterranean Sea by Oil and Other Harmful Substances in Cases of Emergency creates a coordinated assistance program in case of a spill of oil or toxic chemicals.

The 1976 Barcelona Convention entered into force rapidly. By 1978 six countries completed the ratification process, and the framework convention and protocols entered into force on February 12, 1978. Notably, the framework convention designated UNEP to act as the coordinating agency for the MAP, with initial funding coming from the UN Environment Fund. Yeroulanos [56] credited the MAP's success to a small but effective secretariat supporting the signatory countries.

Over the next thirty years, states updated the Med Plan as information about environmental threats became known, and as the political alignments of nations changed with the breakup of the Soviet Union. States funded the RSP through the Mediterranean Trust Fund with UNEP acting as the fund administrator. States added five protocols to the MAP: the Protocol for the Protection of the Mediterranean Sea against Pollution from Land-Based Sources, the Specially Protected Areas and Biological Diversity Protocol, the Protocol for the Protection of the Mediterranean Sea against Pollution Resulting from the Exploration of the Continental Shelf and the Seabed and its Subsoil, The Protocol on the Prevention of Pollution of the Mediterranean Sea by Transboundary Movements of Hazardous Wastes and their Disposal, and the Protocol on Integrated Coastal Zone Management in the Mediterranean.

States also amended the Barcelona Convention in 1995 and changed the name to the Convention for the Protection of the Marine Environment and the Coastal Region of the Mediterranean. This amendment entered into force in 2004. UNEP remained the MAP coordinator with its headquarters in Athens, Greece. Other cities hosting part of the infrastructure to implement this RSP include Valletta, Malta, Marseille, France, Split, Croatia, Tunis, Tunisia, Barcelona, Spain, and Rome, Italy.

The MAP represented the first and most successful of the RSPs. States normally in conflict put their differences aside to protect an essential common pool resource in the Mediterranean. This cooperation is noteworthy given the states involved. Additionally, the agreements improved the water quality in the Mediterranean region as states implemented controls on pollution sources, including industrial facilities and municipal sewage plants.

Tolba [46] stated that the Barcelona Convention's success immediately led to the opportunity to negotiate five other regional seas agreements. As the success of the MAP became apparent, other countries sought inclusion in the program. By 2021, eighteen regional seas conventions and action plans existed. UNEP administers RSPs in the Mediterranean Sea, the Wider Caribbean Sea, the Western and Central African Seas, the Eastern Africa Seas, the East Asian Seas, and the North-West Pacific. Seven RSPs utilize a different regional organization for administration, including the Regional Organization for the Protection of the Marine Environment Sea Area formally known as the Kuwait Action Plan Region, the South-East Pacific, the North-East Pacific, the Red Sea, and the Gulf of Aden, the South Pacific, the Black Sea, and the South Asian Seas.

The five remaining RSPs originated outside UNEP oversight. These programs contributed and continue to contribute to managing and protecting their marine environment. These RSPs cover the Baltic Sea, the North-East Atlantic, the Caspian Sea, the Arctic, and the Antarctic. The Convention on the Protection of the Marine Environment of the Baltic Sea Area (Helsinki Convention) opened for signature in 1974, before the 1975 Barcelona Convention meeting. However, the Med Plan completed ratification before the Helsinki Convention.

4.5 Regimes

The beginning of this chapter identified environmental problems that led to the proliferation of international and regional environmental agreements, each with its own set of rules, participants, and political dynamics. The issues reviewed in this chapter may be conceptualized as common pool resources subject to the tragedy of the commons. Sigvaldsson [38] stated that whaling under the IWC fit into the common pool resource framework.

Given the importance of these problems not only in environmental affairs but also within international diplomacy more broadly, providing a model to analyze these differing environmental agreements moved to the forefront of scholarly inquiry. Departing sharply from the realist arguments about the competition of states, scholars investigating environmental affairs instead focused on the reasons why states cooperate to achieve common goals.

Scholars developed the concept of regime theory to explain why states cooperate, a preferred technique for managing a common pool resource such as whales, fishes, regional seas, and the high seas. Krasner [21] defined a regime as an issue area where actors follow the rules, regulations, and guidelines established by the group in a predictable fashion. International treaties, and the social structures they established, became synonymous with regimes. Regimes developed around singular issue areas with unique characteristics such as underlying problems, institutional rules, and influential actors.

Krasner [21] differentiated between principles, rules, decision-making procedures, and underlying norms, values, and beliefs. Changes within the principles rules and decision-making procedures can be considered changes from within the regime while alterations in the underlying norms, values, and beliefs indicate that the regime is changing.

Sigvaldsson [38] investigated the causes of change within the IWC regime. Sigvaldsson tested five models of regime change, including economic processes, internal contradictions, contemporary power structures, the impact of international organizations, and bargaining and coalition making. He determined that each of the five models contributed to scholarly understanding of regime change and recommended that scholars utilize them in tandem.

Breitmeier et al. [3] pointed out that regimes proliferated to meet states' demand for governance. Once established within the international realm, environmental regimes multiplied as the number of issues upon which states could cooperate to protect the environment became apparent. This chapter reviewed three distinctive regimes—the moratorium on whaling, UNCLOS III, and the Mediterranean RSP. In these cases, the international diplomatic system yielded useful results in institutional capacity building, even if the newly designed regimes did not quite live up to their promise of functioning as designed or delivering on promises of environmental protection.

Regimes assist in ordering analyses about singular issue areas. Within international environmental affairs, each individual regime functions differently than the others. One major line of inquiry with regime theory focused on how and why regimes form. Young [57] suggested that regimes form through a three-phase process, including agenda formation, negotiation, and operationalization. He further detailed that each stage features unique requirements for regimes to develop completely.

Of particular interest, here, is the concept of driving forces in each stage of the regime formation process. Utilizing Young's [57] framing of regime formation, epistemic communities, in the case of the Med Plan, impacted the agenda formation of a nascent regime by providing support for the ideas about problems and solutions. For example, NGOs played an unusually significant role in pressuring states to create a ban on whaling [25, 37]. In this instance, environmental NGOs successfully leveraged new scientific knowledge they helped sponsor along with public opinion to influence states' vote on the ban. Other regimes, such as the Mediterranean Sea, emphasized the role of scientists within the negotiating team [12].

Once within the negotiating phase of regime formation, states' interests dominate, but may not be the only factor of importance during this stage [57]. The rules of the UN heavily favor states, up to and including the states' ability to monopolize decision-making on the actual treaty text. A cursory glance at this chapter and the primary and secondary sources utilized to assemble the historical aspects of this chapter reveal very little mention of non-state actors during this time frame once the diplomats begin the process of crafting treaty text.

Young [57] concluded his analysis of driving forces during operationalization by focusing on material aspects of power, particularly economic power. Implementation requires significant funding, not only in developed countries but also in developing countries. One unanswered question from this chapter revolves around why the Mediterranean Sea became the first of the RSP? Many European countries involved had the funding they could have made available to conduct the research necessary to complete the environmental assessment. Did the European countries prioritize the Mediterranean as a return on their investment stemming from UNEP's Environmental Fund contributions? Did UNEP prioritize the Mediterranean because the European countries could provide funding after UNEP withdrew? It seems likely that both answers could have impacted the decision to begin the RSP in the Mediterranean. Regardless of the answer, tracing how money flows has proven more durable than Young presumed. The critique, however, of regimes struggling to predict change remains one of the more important obstacles that theorists have not overcome. This critique, however, does not diminish the value of regime theory to scholars or students. It does, however, indicate room for further development within the field.

References

1. Berzin AA (2008) The truth about Soviet whaling: a memoir. Mar Fish Rev 70(2):1–62
2. Bliss-Guest PA, Keckes S (1982) The regional seas programme of UNEP. Environ Conserv 9(1):43–49
3. Breitmeier H et al (2006) Analyzing international environmental regimes from case study to database. MIT Press, Cambridge
4. Brown S, Fabian LL (1974) Diplomats at sea. Foreign Aff 52(2):301–321
5. Engfeldt L-G (1973) The United Nations and the human environment. Int Organ 27(3):393–412
6. Fawcett JES (1977) So UNCLOS failed—or did it? World Today 33(1):28–34
7. Ferrari M (1983) Of whales and politics. Ambio 12(6):347–348
8. Frakes J (2003) The common heritage of mankind principle and deep seabed, outer space, and Antarctica: will developed and developing nations reach a compromise. Wis Int LJ 21(2):409–434
9. Gambell R (1977) Whale conservation: role of the international whaling commission. Mar Policy 1(4):301–310
10. Gosovic B (1992) The quest for world environmental cooperation: the case of the UN global environment monitoring system. Routledge, London
11. Haas PM (1989) Do regimes matter? Epistemic communities and Mediterranean pollution control. Int Organ 43(3):377–403
12. Haas PM (1990) Saving the Mediterranean: the politics of international environmental cooperation. Columbia University Press, New York
13. Hardin G (1968) The tragedy of the commons: the population problem has no technical solution; it requires a fundamental extension in morality. Science 162(3859):1243–1248
14. Helmer R (1977) Pollutants from land-based sources in the Mediterranean. Ambio 6(6):312–316
15. Holt SJ (1983) The Indian Ocean whale sanctuary. Ambio 12(6):345–347
16. Hulm P (1983) The regional seas program: what fate for UNEP's crown jewels? Ambio 12(1):2–13
17. Ivashchenko YV, Clapham PJ (2014) Too much is never enough: the cautionary tale of Soviet illegal whaling. Mar Fish Rev 76(1–2):1–22

18. IUCN (2021) The IUCN red list of threatened species. version 2021–2. https://www.iucnredlist.org. Accessed Nov 4 2021
19. Koh TTB (1984) Negotiating a new world order for the sea. Va J Int Law 24(4):761–784
20. Koh TTB (1987) The origins of the 1982 Convention on the Law of the Sea. Malaya Law Rev 29(1):1–17
21. Krasner SD (1982) Structural causes and regime consequences: regimes as intervening variables. Int Org 36(2):185–205
22. Lagrange AS (1977) The Barcelona Convention and its protocols. Ambio 6(6):328–332
23. Lebreton L et al (2018) Evidence that the great Pacific garbage patch is rapidly accumulating plastic. Sci Rep-UK 8(1):1–15
24. Levering RB, Levering ML (1999) Citizen action for global change: the Neptune Group and Law of the Sea. Syracuse University Press, Syracuse
25. Mulvaney K (1997) The International Whaling Commission and the role of non-governmental organizations. Georget Environ Law Rev 9(2):347–354
26. Noyes JE (2008) Louis B Sohn and the law of the sea. Willamette J Int Law Dispute Resolut 16(2):238–251
27. Oslund K (2004) Protecting fat mammals or carnivorous humans? Towards an environmental history of whales. Hist Soc Res 29(3):63–81
28. Ostrom E (1990) Governing the commons: the evolution of institutions for collective action. Cambridge University Press, New York
29. Ostrom E et al (1994) Rules, games, and common-pool resources. University of Michigan Press, Ann Arbor
30. Oxman BH (1977) The third United Nations Conference on the Law of the Sea: the 1976 New York sessions. Am J Int Law 71(2):247–269
31. Oxman BH (1978) The third United Nations Conference on the Law of the Sea: the 1977 New York session. Am J Int Law 72(1):57–83
32. Oxman BH (1979) The third United Nations Conference on the Law of the Sea: the seventh session (1978). Am J of Int Law 73(1):1–41
33. Oxman BH (1980) The third United Nations conference on the Law of the Sea: the eighth session (1979). Am J Int Law 74(1):1–47
34. Oxman BH (1981) The third United Nations Conference on the Law of the Sea: the ninth session (1980). Am J Int Law 75(2):211–256
35. Papastavrou V (2019) Turning the tide: 50 years of collaboration for whale and dolphin conservation. WWF https://files.worldwildlife.org/wwfcmsprod/files/Publication/file/4mcqmmkdiv_wwf_turning_the_tide_041019_1_cetaceans_report.pdf?_ga=2.252887819.344573950.1638145954-591693336.1638145954. Accessed 30 Jan 2022
36. Ryan M (1979, August 20) In a dramatic duel at sea, a young conservationist rams a ship to save the whales. People Weekly, Time Publishing 12(8):30
37. Sakaguchi I (2013) The roles of activist NGOs in the development and transformation of IWC regime: the interaction of norms and power. J Environ Stud Sci 3(2):194–208
38. Sigvaldsson H (1996) The International Whaling Commission: the transition from a whaling club to a preservation club. Coop Confl 31(3):311–352
39. Smith-Godfrey S (2016) Defining the blue economy. Marit Aff J Nat Marit Found India 12(1):58–64
40. Stelios S (2010) "Save the Whales" 35th Anniversary. https://www.greenpeace.org/usa/save-the-whales-35th-anniversary/. Accessed on 28 Nov 2021
41. Stevenson JR, Oxman BH (1974) The preparations for the Law of the Sea Conference. Am J Int Law 68(1):1–32
42. Stevenson JR, Oxman BH (1975a) The third United Nations Conference on the Law of the Sea: The 1974 Caracas session. Am J Int Law 69(1):1–30
43. Stevenson JR, Oxman BH (1975b) The Third United Nations Conference on the Law of the Sea: The 1975 Geneva Session. Am J Int Law 69(4): 763–797
44. Stevenson JR, Oxman BH (1994) The future of the United Nations Convention on the Law of the Sea. Am J Int Law 88(3):488–499

45. Thacher PS (1977) The Mediterranean action plan. Ambio 6(6):308–312
46. Tolba M (1977) Introduction. Ambio 6(6):299–299
47. UNCHE (1972) Report on the United Nations Conference on the Human Environment. A/CONF.48/PC11
48. UNCLOS (1982, December 10) Settlements of dispute mechanism. https://www.un.org/depts/los/settlement_of_disputes/choice_procedure.htm. Accessed 9 Jan 2022
49. UNEP (1974) Report of the Governing Council on the work of its second session. A/9625
50. UNEP (1982) Achievements and planned development of UNEP's regional seas program and comparable programs sponsored by other bodies, UNEP Regional Seas Reports and Studies No. 1
51. UN GA (1967) General Assembly—twenty-second session First Committee, 1515 Meeting. A/C.1/PV.1515
52. van Drimmelen B (1991) Comment, the international mismanagement of whaling. UCLA PBLJ 10(1):240–259
53. Ward JE (1982) Game review: save the whales by Kenneth E. Kolsburn. Animal Town Game Co, PO Box 2002, Santa Barbara, CA 93120 Copyright 1978 $17.00 Simul Gaming 13(1):122–124
54. Walsh VM (1999) Illegal whaling for humpbacks by the Soviet Union in the Antarctic, 1947–1972. J Environ Dev 8(3):307–327
55. World Bank and UN DESA (2017) The potential of the blue economy: increasing long-term benefits of the sustainable use of marine resources for small island developing states and coastal least developed countries. World Bank, Washington, D.C.
56. Yeroulanos M (1982) The Mediterranean Action Plan: a success story in international cooperation. Ekistic 49(293):175–179
57. Young OR (1998) Creating regimes: Arctic accords and international government. Cornell University Press, Ithaca
58. Zelko F (2017) Scaling greenpeace: from local activism to global governance. Hist Soc Res 42(2 (160)):318–342
59. Zemsky VA et al (1995) Soviet Antarctic whaling data (1947–1972) Center for Russian Environment, Moscow

Chapter 5
Protecting the Ozone Layer

Abstract International diplomats faced a significant diplomatic challenge when scientists discovered ozone depleting substances had created a hole in the ozone layer. In the absence of complete knowledge about the chemical processes that created the ozone hole, diplomats responded by adapting the international treaty process to create the framework convention and protocol structure that allowed states to move forward. As new scientific discoveries increased the concern about exposure to ultraviolet radiation, states agreed to a phase out of the ozone depleting substances. In doing so, the Vienna Convention for the Protection of the Ozone Layer and its related Montreal Protocol on Substances that Deplete the Ozone Layer became the gold standard for international environmental treaties. Accordingly, scholars generated new theoretical models about the relationship between science and international environmental diplomacy.

Keywords Ozone layer · Montreal Protocol on Substances that Deplete the Ozone Layer · Precautionary principle · Discourse analysis · Vienna Convention for the Protection of the Ozone Layer · Framework convention and protocols

Damage to the environment is not always readily apparent. While some environmental damage may be visible such as oil slicks, other nefarious forms of environmental damage may be invisible, such as air pollution. This invisible killer may cause harmful effects ranging from increasing the frequency of asthma to black lung, an incurable disease caused by inhaling coal dust. Additionally, air pollution may cause human health impacts unrelated to lung disease or the respiratory system. Ozone depletion in the stratosphere causes a greater frequency of skin cancer and cataracts, among other human health impacts. This damage is not limited to humans; other life forms are also at risk. Greater exposure to ultraviolet (UV) radiation also poses severe risks to land-based and water-based life forms, including phytoplankton, an important food source for fish. Crops, in particular, become less productive in more substantial concentrations of UV radiation, threatening the stability of the food supply of humans and animals alike. ˙

Scientists discovered the ozone hole in the latter part of the twentieth century. Realization of the implications of the void became one of the most frightening episodes in modern environmental history. Two research scientists, Drs. Mario Molina and

Frank Sherwood Rowland of the University of California researched compounds suspected of causing ozone loss, particularly chlorofluorocarbons (CFCs). Their research definitively established that CFCs, initially thought to be benign chemicals, caused ozone destruction [19]. This theory, commonly referred to as the Molina–Rowland theory, forever changed our collective understanding of environmental science and launched a search for a greater understanding of the impacts of these chemicals on the natural environment. Twenty years later, in 1995, Molina and Rowland won the Nobel Prize in Chemistry for their work on ozone depletion.

The creation of the Molina–Rowland theory of ozone destruction served as a starting point for a greater understanding of the stratosphere. The scientific process is precisely that; it is a process for discovering new knowledge. This process does not mean that the correct answer appears with each new experiment. For example, the initial ozone models suggested that CFCs caused a significant thinning of the ozone layer [26]. As scientists understood more about the ozone layer, a series of model refinements in the late 1970s and early 1980s suggested that the ozone layer was not thinning as quickly as expected due, in part, to a series of mandatory reductions in the United States and voluntary reductions in Europe and elsewhere around the globe [2].

In May 1984, three British scientists, Joe Farman, Brian Gardiner, and Johnathan Shanklin, realized that ozone over Halley Bay, Antarctica, had dramatically declined in every Antarctic spring season dating back to at least 1981. They discovered that the ozone hole did not occur uniformly everywhere in the stratosphere but instead concentrated at the South Pole. Communicating this loss to the global population increased awareness that significant environmental harm had already occurred.

As a result, public pressure to protect the ozone layer increased. As this astounding news entered public consciousness, an overriding need to heal the stratosphere redirected the efforts of scientists, diplomats, corporations, and ordinary people to eliminate products utilizing the damaging chemicals [16]. Scientists scrambled to understand the cause of the ozone hole in hopes of reversing the damage quickly. Diplomats worldwide mobilized as the need for swift, decisive action to negotiate a halt to the use of the chemicals causing the ozone hole became apparent. Corporations, both those that produced the ozone-depleting substances and corporations that used the ozone-depleting substances, launched a search for economically efficient and environmentally friendly substitutes [2]. In the meantime, ordinary citizens mobilized to boycott products that used ozone-depleting substances and pressure governments to take decisive actions by enacting domestic laws and creating international treaties.

Today, science accepts that the stratospheric circulation results in a "polar vortex" over Antarctica. Anthropogenic chemicals in the stratosphere build up during the winter months, increasing the photochemical depletion of ozone during early Spring. This buildup creates the ozone hole, typically during September–November over the South Pole in Antarctica.

The European Environmental Agency and the National Aeronautics and Space Administration (NASA) released and continue to release photos of the ozone hole as a series of color-coded circles around the South Pole [9, 22]. After discovering the ozone hole, the size and duration of the ozone hole over Antarctica became the defining symbol of this environmental problem. These images demonstrate the size and shape of the changes in the ozone hole over time [9, 22].

Like so many other episodes within international environmental politics, this story is a story of the many thousands of individuals who took action to stop and reverse environmental damage. Like previous chapters, the ozone story begins with a brief explanation of atmospheric science focusing on the cause and effect of the ozone hole. The second section reviews the diplomatic processes and the innovations in negotiating techniques that revitalized the international environmental agenda. The third section focuses on understanding the new models produced by academics working in the field. The fourth and final section reviews the changes in ozone diplomacy after negotiations end.

5.1 From Science to Vienna

Ozone is a naturally occurring substance in the stratosphere, a layer within the upper atmosphere approximately 9 to 18 miles above the earth's surface. Ozone within the stratosphere is concentrated in what today we call the ozone layer that absorbs UV radiation, including UV-C that is lethal to all forms of life. The ozone layer also blocks UV-B, harmful but not immediately lethal radiation, from reaching the surface.

The relationship between the thickness of the ozone layer and damaging UV radiation is at once, both simple and complex. As the ozone within the band thickens, less UV radiation reaches the earth. Conversely, as the ozone within the band thins, more UV radiation reaches the ground. Thus, measuring the thickness of the ozone band within the stratosphere came to represent the health of the ozone layer. Consequently, science, and the scientists that perform the experiments, became intricately linked to the negotiating process as the thickness of the ozone band was initially the primary measurement used to justify or delay calls for more state cooperation in limiting ozone-depleting substances [16].

Complicated science enters the equation because the chemical reactions that maintain the thickness of the ozone layer require a delicate balance so that ozone is neither created nor destroyed. However, this balance is fragile. Human activities, such as chemical usage at the earth's surface, may alter the chemical reactions that keep the amount of ozone in the ozone layer constant. Ozone-depleting chemicals reach the stratosphere by floating up through the troposphere (lower atmosphere). Once these CFCs react in the presence of naturally occurring UV radiation to break off a free chlorine. This free chlorine then interferes with the typically balanced ozone by causing ozone to break apart faster than it naturally reforms.

Chemicals such as hydroxyl, nitric oxide, chlorine, and bromine catalyze the chemical reactions that destroy ozone. Thus, classes of gasses such as CFCs, halons, hydrochlorofluorocarbons (HCFCs), carbon tetrachloride, methyl bromide, hydrobromofluorocarbons (HBFCs), chlorobromomethane, and methyl chloroform all contribute to the thinning of the ozone layer. When the free chlorine finishes destroying an individual ozone molecule, the free chlorine attacks another ozone molecule. In other words, every free chlorine in the stratosphere attacks and destroys several hundred ozone molecules. The result is that adding these catalytic chemicals into the atmosphere destroyed the ozone in the ozone layer faster than it could be replaced.

Simply plugging the ozone hole by adding ozone to the upper atmosphere, perhaps as a type of ozone immunization booster, does not show promise as a geoengineering solution as ozone does not move easily from the troposphere to the stratosphere. Ending atmospheric releases of catalyst chemicals that alter the chemical balance of the stratosphere may be the only way to end the ozone hole in the future as we also currently do not have the ability to remove the catalyst chemicals that are already present in the stratosphere [24].

The scientific process, however, is neither failproof nor straightforward. Doolittle [5] believed that scientists continuously underestimated the dangers of stratospheric ozone destruction. The scientific discovery process made multiple missteps and miscalculations. One of the most flawed decisions regarding the systemic study of the ozone layer occurred when a group of American scientists disregarded all ozone measurement data that showed a reduction of more than 30% of ozone [15]. This computer programming failure occurred in the 1970s but was only discovered after Farman et al. [10] published their results. This programming failure caused a delay in detecting the hole in the ozone layer and also delayed the actions necessary to reverse this damage.

A critical lesson from this unfortunate circumstance is a noticeable difference between understanding the causal mechanism by which chlorine in the stratosphere destroys ozone and accurately predicting the impact of the destruction of ozone upon human life and the natural environment. Theoretically, understanding how environmental damage occurs may not translate into successfully creating predictive modeling for understanding environmental impacts.

Predictive modeling became deeply embedded in the political processes as a "best guess" may be better than no information consequently [14]. Thus, the decision-making process may be based partially upon the clarity of the science, including modeling. Actors within international diplomacy consistently decide whether to act now or wait until later based upon the probability of worse environmental damage if they wait until later. This decision is not a once-in-a-lifetime choice but rather a routine part of the negotiating process and is represented by the precautionary principle. The precautionary principle suggests that it is better to act in the face of scientific uncertainty than to wait and fatally damage the environment's ability to support life in the future.

Chemical corporations rarely favor the precautionary principle as it frequently requires a change in operations that decreases profits. Further, the precautionary principle may require ending production lines in their entirety. Corporations thus balance competing requirements between making profits for shareholders and engaging in activities that protect the public and the environment. In this environmental arena, The Dow Chemical Company, ostensibly the largest producer of CFCs and other chemical manufacturers such as DuPont and the Imperial Chemical Industries, used

CFCs, along with other ozone-depleting substances, to make commercially available products such as refrigerants, aerosols, and plastic foam, including insulation [2].

Makhijani and Gurney [17] produced an excellent treatment of how these corporations' products may be used in other industries. Products that utilize the CFCs may have trivial uses within society, such as the propellant for hair spray. However, other services are more significant, such as refrigeration in automobiles and housing insulation that protects occupants from extreme cold. Halons may be found in necessary fire protection in airplanes and marine fairing vessels. For many years, asthma inhalers also utilized ozone-depleting substances as the propellant.

Consequently, eliminating all ozone-depleting chemicals across the board may not serve the best interests of all of humanity. However, continuing trivial commercial uses of these ozone-depleting substances, such as propellants in the dispersal of hair spray, could not be justified, given the potential for such catastrophic loss of life. In response to the immediate threat to their profits, major chemical companies began searching for alternative products that could replace these chemicals while lowering their potential to deplete the ozone in the ozone layer. Thus, multinational corporations became an essential voice within international negotiations on protecting the ozone layer [2, 5, 16].

After the acceptance of the Molina and Rowland theory, national governments needed to decide what, if anything, they were willing to do about stratospheric ozone depletion, including halting multinational corporations' production of damaging products within their domestic territory. UNEP, exemplifying its role as a coordinator and incubator of international environmental leadership, convened a meeting of experts in 1977 to consolidate and coordinate the current state of the science regarding ozone depletion. This group produced the World Plan of Action on the Ozone Layer containing a 21-point research plan.

The World Plan of Action on the Ozone Layer furthered the process of monitoring data to determine the extent to which the ozone layer had been damaged by requesting that international organizations coordinate scientific surveys to gather data about the amount of damage to the environment. Scientists also collected and disseminated data about the potential impact on human health and appraised the costs and benefits of eliminating the destruction of the ozone layer. Further, UNEP created the Coordinating Committee on the Ozone Layer that consisted of national governments, international organizations, and NGOs to implement the World Plan of Action. The committee assessed the current status of ozone science and recommended further actions to reduce ozone depletion based upon the best available knowledge at the time. The Coordinating Committee on the Ozone Layer, made up predominantly of states, also included a handful of NGOs. Typically, the committee met once a year and became one of the primary focal points of ozone diplomacy. Negotiators looked to this group to provide guidance on the science behind the ozone hole [2].

While NGOs did participate in the Coordinating Committee on the Ozone Layer, there is no evidence to suggest the group impacted the outcome. Participants' lists from this series of meetings indicate that some sessions did not have an environmental NGO participant. However, the Chemical Manufacturer's Association did participate on behalf of business and industry. This difference in access indicated that business and industry groups had more opportunities to influence the output of this group vis-à-vis other more environmentally conscious NGOs. Additionally, allowing business and industry groups to join the committee meant that these companies could allow scientists to analyze global production data as part of the scope of work for this group. As a result, this group correlated the thinning of the ozone hole to product usage and increases in chemical production.

In the ozone arena, the United States took a lead role to limit damage to the stratosphere by banning CFC usages of aerosols in late December 1978 under the Toxic Substances Control Act (TSCA). The EPA, the Consumer Product Safety Commission, and the Food and Drug Administration acted in concert to create the ban out of concern that aerosols may negatively impact the ozone layer [7].

Nor was the United States the only country to issue regulations dealing with the trivial use of CFCs. By 1981 consensus emerged from the scientific community that continued use of CFCs in non-essential aerosols such as cosmetics, deodorants, and hair sprays should end. Individual countries, including Canada and the Netherlands, joined the United States in enacting regulations that limited the use of CFCs. While concerned about the continued use of CFCs in aerosols, other countries in Western Europe either took voluntary action or waited for regulatory action through the Council of the European Economic Community. In an ironic role reversal compared to contemporary environmental politics, the United States became the primary champion of this environmental treaty. The European countries did not favor internationally binding commitments.

Notably, the total of the mandatory and voluntary agreements did not permanently limit the overall production of CFCs, as the decline in CFC use in aerosols was partially offset by the growth of CFCs in refrigeration, including automobiles and homes. This situation led to the possibility that while chemical manufactures made and sold fewer CFCs, this decrease in production and sales might not continue in the future. It also meant that ozone-depleting substances continued to accumulate in the stratosphere due to their longevity.

In 1981, UNEP included work on protecting the ozone layer as part of the Montevideo Law Program, a work plan intended to articulate a broad vision of environmental law as a follow up to the 1972 UNCHE conference [29]. During this meeting, Finland presented a draft of a global framework to further discuss the ozone layer's protection. This draft would serve as the basis for the Vienna Convention for the Protection of the Ozone Layer (Vienna Convention). Andersen and Sarma [2] report that the draft contained the key elements of the resulting treaty: sharing scientific information, using science as the basis for future negotiations, developing new technologies, control policies, and reduction strategies as appropriate, and establishing an international convention implemented by states.

UNEP's Governing Council, at its ninth session in May 1981, authorized work to create a convention for the ozone layer's protection [28]. Consequently, UNEP convened the Ad Hoc Working Group of Legal and Technical Experts for Elaboration of a Global Framework Convention. This group convened in Stockholm, Sweden, in January 1982. The initial Finnish paper became the official draft text of the meeting.

Structurally, the draft included details on countries' participation in the treaty negotiating processes, the time scale for future meetings, and the nomination of UNEP as the conference secretariat. The draft also included the establishment of a science and technology committee, a general agreement to share information about ozone depletion, and a mandate to prevent activities that further contribute to damaging the ozone layer [2].

Countries negotiating the Vienna Convention split into two camps. The first bloc, the Toronto group, consisted of the United States, Canada, Finland, Norway, and Sweden. This group sought binding mandatory reductions of all CFCs as part of the Vienna Convention. In opposition to the Toronto Group, the European Economic Community, the Soviet Union, and Japan sought instead to postpone the determination of mandatory reductions of CFCs until science could provide definitive estimates of the reductions needed to avoid damaging the ozone layer. The main portion of the treaty—a control schedule for ending the use of ozone-depleting substances, is noticeably absent from this initial draft treaty.

5.2 From Vienna to Montreal

States met formally to discuss the Vienna Convention from March 18–22, 1985. Compared to past international environmental meetings, the participants completed negotiations in a short time frame. Only thirty-four countries were in attendance, with a scant handful of NGOs. Perhaps the most prominent of these NGOs was the International Chamber of Commerce. Equally telling, no environmental NGOs, such as Greenpeace, attended the meeting.

The Vienna Convention detailed agreements among the states to continue researching ozone depletion, including information sharing regarding scientific observations and cooperating to create policies that might limit human impact on the ozone layer [32]. The Vienna Convention did not include any legally binding commitments to reduce emissions or production capabilities for ozone-depleting substances as national governments' differences proved too challenging to overcome. As the Toronto Group and the European Economic Commission could not reconcile their differences. Committee delegates agreed to accept UNEPs offer to house the convention secretariat while the parties to the Vienna Convention decided on a permanent home for the secretariat [32]. These countries also agreed to partially fund secretariat activities, including contributing to some of the initial costs. These actions signaled that UNEP functioned as designed as a catalyst of international environmental affairs.

Notably, the Vienna Convention parties agreed to continue the conversation on creating a specific protocol later, should the individual states decide that the ozone science required further action [32]. This action represented a slight departure from the RSP, as discussed in Chap. 3. Recall that in the Mediterranean RSP, states negotiated the framework convention and the underlying protocol simultaneously. Here, states negotiated the framework convention first and agreed to create additional protocols as new scientific information became available.

The Vienna Convention requires that a Conference of the Parties (COP) occur at fixed intervals; today, the COP meets once every three years.[1] The Vienna Convention declares that the first COP should occur no more than a year after entry into force of the Convention. Further, the Vienna Convention required the Executive Director of UNEP to begin work on a protocol to be completed, if at all possible, by 1987. Thus, UNEP remained the conference organizer while seeking to reconcile the all-important decision about binding commitments to limit ozone-depleting substances.

An assumption within international environmental politics emerged that all UN member states participated in negotiating the Vienna Convention under the watchful eye of a dedicated and energized civil society. Nothing could have been further from the truth. The official rosters of the Vienna Convention included a scant 34 member-states (out of 159 member-states at the time) and three NGOs, all representing business interests. Countries were allowed to ratify the Vienna Convention for 12 months after it opened for signature. After that date, countries were allowed to accede to the treaty but not ratify the treaty.

UNEP convened a second working group, the Ad Hoc Working Group of Legal and Technical Experts for the Preparation of the Protocol on Chlorofluorocarbons to the Vienna Convention for the Protection of the Ozone Layer that worked toward creating the Protocol for the Vienna Convention. Under the terms and conditions of the Vienna Convention, the meetings were convened by the Executive Director of UNEP, Mostafa Tolba from Egypt. This committee focused on four key issues: scientific and technical issues specific to developing countries, regulatory measures, and trade [30]. As is often the case with complex environmental negotiations, a consensus emerged slowly and in a piecemeal fashion. Consensus items included restricted imports from non-parties to the protocol and that the financing and administration of the protocol should come from funds contributed from the parties to the protocol.

Further testimony from scientists ramped up the pressure for governments to finalize the protocol. Perhaps the most politically salient testimony came once again from Dr. F. Sherwood Rowland when he testified in front of the United States Congress in 1987 that not only did CFC cause ozone depletion over Antarctica but also over the United States [14]. Consequently, the United States hardened its negotiating position, insisting on more rigid controls and a shorter time frame for implementation.

Industry groups such as the Alliance for Responsible CFC Policy reversed course, dropping its opposition to an internationally negotiated protocol with legally binding emission reductions and signaling industry acceptance of binding targets. In part, chemical companies realized that retooling to produce the replacement chemicals

---

[1] Many conventions convene a COP after a treaty has entered into force. Diplomatic tradition numbers the meetings in order by topic. This tradition is utilized in the text. Consequently, there may be references to COP–1 under the Vienna Convention and COP–1 under the UNFCCC, referring to two separate meetings.

would not significantly impact their profits. Additionally, public opinion in the United States shifted negotiating positions based on expanding scientific knowledge that concretely linked these CFCs to the ozone hole [16]. As a result of these two shifts from prominent actors within the negotiating processes, other governments responded in favor of banning CFCs.

Benedick [3] and Andersen and Sarma [2] credited UNEP's Executive Director Mostafa Tolba with the political leadership that created the breakthrough. His negotiating skills allowed the final text of the Montreal Protocol to emerge at the third session of the Working Group, held in Geneva from April 27 to 30, 1987. Tolba's commitment to informal consultations and formal meetings allowed for the creation of an informal draft text on control measures that states would continue to refine over the summer of 1987.

The COP in Montreal was not a large gathering. A scant fifty-five countries attended as participants, with seventeen NGOs in attendance. For comparison purposes, UNEP's Governing Council contained fifty-eight member states. These NGOs predominantly represented business and industry groups, particularly chemical manufacturing companies that sought to protect their commercial interests. Chemical manufacturers initially sought to continue producing products made from ozone-depleting substances (or their substitutions). Additionally, six intergovernmental organizations participated, representing other agencies within the UN system and different intergovernmental organizations in attendance.

The Montreal Protocol recognized that precautionary action was needed to protect the ozone layer and thus contained a schedule for a phase-out of the production and the consumption of certain CFCs and halons. Keeping in mind that the diplomatic process remains a political negotiation, albeit one guided and informed by science, diplomats listed some, but not all, of the ozone-depleting substances in Annex A and split the chemicals into two groups. Group I consisted of CFCs subject to a phase-out schedule, while Group II included halons subject to a production freeze. Developed countries took the lead in eliminating ozone-depleting substances as developing countries phase out deadline occurs ten years after the date for developed countries. Group I chemicals typically have lower ozone-depleting potentials than Group II chemicals; however, consumption of Group I chemicals outweighs the consumption of Group II chemicals. Due to their combined impact, Group I chemicals must be phased out earlier than Group II chemicals. Group II chemical production will be banned beginning no later than January 1, 2030.

Further, developing countries requested that developed countries provide the necessary funding to adopt substitution chemicals and meet compliance obligations under the Montreal Protocol. Equally importantly, the Montreal Protocol contained provisions to reevaluate the production and use ban based on recent science. Provisions were then inserted to ensure that parties to the protocol considered new scientific discoveries at least once every four years, beginning in 1990.

States finalized the Montreal Protocol on September 16, 1987. Shortly after the Montreal Protocol opened for signature, environmental groups organized into networks to create domestic support for ratification. This domestic support would pressure governments to sign the Vienna Convention and the Montreal Protocol

[4]. With its network of like-minded groups, Friends of the Earth launched campaigns aimed at national governments and corporations that used ozone-depleting substances. Within the United States, Friends of the Earth launched the "Styro-Wars" campaign aimed at styrofoam products, including cups distributed at fast-food chains like McDonald's or Wendy's [23].

The Meeting of the Parties (MOP) that ratified the Montreal Protocol convenes once a year using the guidelines set out in the Montreal Protocol. Like the COP to the Vienna Convention, the Ozone Secretariat, based out of the UNEP, provides the MOP with logistical support. While some meetings were low-key and technical, other diplomatic meetings sought to enact considerable changes to the ozone regime. Important topics included quickening the phase-out of CFCs and other ozone-depleting substances and adding chemicals to the list of banned substances. Scientific discoveries after negotiating the Montreal Protocol confirmed the worst fears of the environmental groups; the number of CFCs and halons already in existence would significantly damage the ozone layer after considering the limitations on production contained in the Montreal Protocol [13]. Consequently, parties to the Montreal Protocol continue to amend the treaty periodically, with crucial amendments occurring in 1990 (London Amendment), 1992 (Copenhagen Amendment), 1997 (Montreal Amendment), 1999 (Beijing Amendment), and 2016 (Kigali Amendment).

The 1990 London Amendment altered the Montreal Protocol by strengthening controls on the original CFCs and halons and adding new chemicals to the Annex subject to controls. Diplomats added ten CFCs, carbon tetrachloride, and methyl chloroform to Annex A chemicals subject to phase-outs on production and consumption. Additionally, the London Amendment added a Multilateral Fund that assists the 147 Article 5 developing countries whose annual consumption of CFCs and halons is less than 0.3 hg per capita. The fund has received over USD 4.07 billion since its inception in December 2019 to implement projects and activities that reduce emissions, including the phase-out of CFCs, halons, and HCFCs [21]. Montreal, Canada, hosts the Multilateral Fund Secretariat. Donor countries, i.e., the developed countries, replenish funds every three years.

Like the London Amendment, the Copenhagen Amendment widened and deepened the phase-out schedule for ozone-depleting substances. The Copenhagen Amendment adds HCFCs and methyl bromide to the phase-out schedule. Additionally, diplomats agreed to add HBFCs to the phase-out schedule. These chemicals do not have a significant commercial use, with the possible exception of HBFC-31 that is used in pharmaceutical manufacturing. Thus, for many HBFCs producers, this addition to the phase-out schedule bans future commercial manufacturing.

Returning to Montreal, Canada, in 1997, MOP-9 to the Montreal Protocol once again updated the phase-out schedule for ozone-depleting substances. States banned the trade of methyl bromide, a common fumigant and pesticide. Additionally, a new action item emerged on the ozone-depleting substances agenda—halting the international trade of banned substances. UNEP [31] notes that the global black market emerged in the mid-1990s due to the pricing disparities between CFCs and their substitute chemicals. Spurgeon [27] notes that freon for use in automobile refrigeration was commonly smuggled into the United States. In the value of merchandise recovered along the United States–Mexico border, freon ranked second in value only to marijuana for the 1994–1996 time period. CFCs cost less, and a considerable amount of older equipment, including automobiles, could not be easily converted to CFC substitutes. Perhaps unsurprisingly, Asia emerged as a hot spot for what has sometimes been referred to as "CFC smuggling." The MOP responded by adopting a licensing system for the import and export of banned and controlled substances.

At MOP-11, held in Beijing, China, the delegates agreed that bromochloromethane should be considered a controlled substance subject to phase out and control measures, specified new production limits on HCFCs, and new restrictions on the trade of these chemicals with non-parties to the Montreal Protocol. When combined, these actions allow for greater environmental protection with an end goal to close the ozone hole more rapidly.

With the realization that certain ozone-depleting substances might also contribute to global warming leading to climate change, parties to the Montreal Protocol made a strategic decision to negotiate further controls on hydrofluorocarbons (HFCs) under the Vienna Convention rather than utilizing the diplomatic processes established by the UNFCCC (see Chaps. 8 and 13). Thus, states chose the diplomatic framework most likely to lead to stronger regulations of these chemicals. Diplomatic proposals to regulate the phase-out of CFCs first began to circulate in 2009, but this work did not come to fruition until the Kigali Amendment in 2016.

The Kigali Amendment specifies the phasedown of HFCs, one of the chemical substitutes for ozone-depleting CFCs and HCFCs. Diplomats seeking to prevent climate change chose to limit the HFCs under the Montreal Protocol to avoid the deadlock that ensnared the climate change negotiations. International diplomats decided to limit HFCs as part of the ozone regime using the justification that HFCs' market share grew due to their usefulness in replacing CFCs and HCFCs. Under the Kigali Amendment, signatories agreed to gradually reduce their HFCs emissions by approximately 80 percent no later than 2047. The Kigali Amendment entered into force on January 1, 2019, after receiving 65 signatories. However, major industrial countries such as the United States have not ratified the Kigali Amendment, severely limiting the potential usefulness of this attempt to limit climate change.

5.3 The Gold Standard of Treaties

The Vienna Convention rarely makes headlines as a watershed moment in the history of international environmental diplomacy. The Vienna Convention certainly did not have the widespread support of both developing and developed countries at the time during which the treaty was negotiated. Doolitte [5] points out that the Vienna Convention could have gone down in the history books as a historical oddity as there was no guarantee that states would attempt to solve other transboundary problems in the same manner. The Vienna Convention was one of the first conventions that attempted to establish solutions to an environmental problem that was inherently transboundary. In other words, the environmental harm moves easily across national borders. Action by one country acting on its own would be wholly insufficient to solve the environmental problem.

The Vienna Convention is also noteworthy for its departure from a single international treaty containing binding commitments to the so-called framework-protocol structure. The Vienna Convention kept the negotiations alive at a critical juncture that allowed time for the science to develop. Thus, the Montreal Protocol garners attention for its novel contributions to protecting life on Earth. Litfin [16] points out that the most critical content of the Vienna Convention is the underlying norm that states have an obligation to refrain from taking actions that damage the environment. Sand [26] argues instead that it is the national government's decision to enshrine an intergenerational element of environmental protection that will be the lasting contribution to international environmental law.

Today, the Montreal Protocol on Substances that Deplete the Ozone Layer serves as the pinnacle of success for international environmental diplomacy. Former UN Secretary-General Kofi Annan went so far as to proclaim the Montreal Protocol as the most successful of the international environmental treaties [1]. In the immediate aftermath of the signing of the Montreal Protocol, many have proclaimed the framework-protocol structure of the Montreal Protocol as a gold standard within international environmental politics [8, 11]. Indeed, the Montreal Protocol remains the sole international environmental treaty that every country in the world has signed and ratified. However, universal ratification was not instantaneous, with the last ratification deposited in 2009.

During much of this time frame, United States Ambassador Richard Benedick reflected on his experiences and the changes to the structure of international environmental treaties [3]. He proclaimed the superiority of this international environmental treaty and stated nine lessons that he believes other environmental negotiating processes should follow. These nine lessons included: enhancing the role of the scientific community, acting with wisdom in the face of uncertainty, the assistance of a well-informed public opinion pressuring both governments and individual corporations to act in the public interest, using multilateral diplomacy to craft a treaty capable of reversing environmental damage, the need for a major country investing in the necessary political bargaining to create diplomatic consensus, lowering barriers to international treaty terms and conditions by creating domestic legislation, utilizing the skills of civil society, structuring treaties to take into account different economic and structural considerations, and working with markets to stimulate necessary technological changes [3]. These nine lessons will significantly shape the climate change negotiations as groups both for (and against) future environmental treaties might attempt to manipulate one or more of these conditions to replicate the "gold standard" in international environmental diplomacy. Environmentalists will seek to keep these things constant; opposition forces will seek to block one or more of these conditions from occurring. In other words, Benedick suggests that one way to defeat the climate change treaty is to prevent a uniform well-informed public opinion, thus giving rise to the climate skeptics that debate whether climate change is currently occurring.

Despite its standing as the most significant international environmental treaty, it is essential to note that in the process of negotiating the Montreal Protocol, more went right than wrong. The expectation that the same set of circumstances will emerge in every situation afterward may well be a type of environmental Pollyannaism that may inadvertently delay more meaningful environmental protection and thus doom individuals and societies to suffer more environmental harm than might otherwise be the case.

Scholars point out various rationales for the Montreal Protocol's success. Patchell and Hayter [25] noted that while the Montreal Protocol utilized the existing state-centric architecture of the UN in that states would regulate their domestic entities, the bulk of the compliance burden fell onto a limited number of corporations conveniently grouped in the same country. Consequently, the Montreal Protocol asked governments to take an action that these same governments were well equipped to carry out in that they are creating new laws that impact few domestic interests. Thus, this action did not involve any surrender of sovereignty, wealth, or technological know-how.

Benedick [3] claimed that the scientific understanding of the ozone hole did not directly impact negotiations in Montreal in 1987 as some of the most revolutionary concepts regarding the ozone destruction mechanism were published after the talks ended. The framework-protocol structure allowed negotiations to continue in the face of scientific uncertainty. As with many other environmental problems, environmental science surrounding the ozone hole remains complex and partially unknown. Diplomats solved this dilemma by embedding science, but not necessarily individual scientists, deeply into the architectural structure of the Vienna Convention and the Montreal Protocol. The scientific discoveries that identified the cause of the damage motivated governments to move forward with negotiations and suggested solutions that could, in time, close the ozone hole.

The relatively low number of corporations impacted by the regulations created a pathway for less resistance. Chemical companies switched production from CFCs to HFCs that caused less damage to the environment, even if the HFCs cost more to produce [20]. As a general rule, the Montreal Protocol allows the ability to substitute ozone-depleting chemicals that cause less damage to the ozone layer for ozone-depleting chemicals that cause more damage, assuming that the substituted chemical is itself not banned. This substitution pathway gave impacted chemical manufacturers the ability to continue to make a profit that also may help explain the success of the Montreal Protocol.

5.4 New Models Emerge

As the methods of conducting international environmental treaties became more complex and varied, scholarly assessment resulted in multiple new insights into the international environmental negotiating processes. Scholars scrutinized the various players within treaty negotiations, such as national governments, international organizations, NGOs, scientists, and businesses [3, 16, 20]. Additional research focused on the role of science and key concepts such as compliance and enforcement. As these changes in international environmental diplomacy became deeply embedded in practice, the academic theories used to analyze international environmental diplomacy became a compelling component of the academic literature.

While the expansion of actors that academics analyze began with the RSPs in Chap. 4, more scholarship focused on the role of non-state actors during and after the ozone layer negotiations. First, Benedick [3] pointed out the change in the role of the environmental NGOs. These groups assumed two essential functions: mobilizing support from the general public and pressuring states for deeper environmental protection. Benedick believed that environmental NGOs possessed more influence when they demonstrated a strong understanding of technical details and avoided emotional-based arguments or political grandstanding. Second, Litfin [16] proposed the discursive practices model, suggesting that it is not the role of the scientific community that is important, but rather the role of science in general. A further explanation of these two models follows.

Benedick [3] stated that one of the pivotal moments that changed the structure of international environmental negotiations became visible at the 1989 London Conference on Saving the Ozone Layer, hosted by the United Kingdom's Prime Minister Margaret Thatcher. This moment highlighted the strength of the environmental NGOs' networks and their ability to interact with the global public. As a result of this meeting, environmental NGOs became more involved in the ozone negotiating process.

The scholarly models reflecting international environmental negotiations added a robust new conceptualization equally important to this field. Litfin [16] extensively analyzed the contributions of the scientific community and concluded that Haas [12] theory of epistemic communities could not be applied to this case. Litfin points to scientists' unwillingness to craft public policy as an essential part of her argument for using a discursive approach. She believes that the role of science as a process for creating new knowledge has more explanatory power than Haas' theory of epistemic communities.

Litfin points out that scientists may lose control over the framing of that knowledge once knowledge has been created. Other political entities that she calls knowledge brokers may use that knowledge for political purposes. Thus, entities that have the ability to control or manipulate knowledge may do so, in turn limiting the diplomatic conversation, including the range of policy options that might come under consideration. As such, discursive practices, the shaping of meaning and context, do not rely on any particular individual or organization to foster institutional change.

5.5 What's Next?

Scientific monitoring of the ozone hole as it heals continues to be sponsored by national governments using various monitoring equipment and satellite imagery [9, 22]. Over time, the size of the ozone hole and the length of time for which the ozone hole is present is shrinking. Aided by upgraded technological sophistication, atmospheric scientists predict the ozone hole will close much more quickly than envisioned in the late twentieth century, provided every country continues to follow the phase-out schedules for all of the chemicals contained in the Montreal Protocol as amended.

However, the detailed modeling of the ozone layer also created evidence of a new phenomenon, CFC smuggling, more formally known as the illegal trade in ozone-depleting substances. The restrictions on the production and use of ozone-depleting substances, but not sales of ozone-depleting substances, gave rise to a black market for these same substances, including CFCs [6]. Profit opportunities arose due to the gap in costs between the now illegal chemicals and their more expensive replacements.

With the decrease in atmospheric emissions of ozone-depleting substances and the increases in both technological monitoring systems and a greater understanding of the functionality of the upper atmosphere, scientists are now able to monitor the quantity of ozone-depleting substances emitted and predict a likely location for their origin. For example, in May 2019, the British Broadcasting Company (BBC) reported that atmospheric scientists discovered previously that the ozone hole was not closing as fast as expected due to a leveling off of the decrease in CFC-11 emissions that should have ended under the Montreal Protocol [18]. In other words, CFC-11 production did not end when it should have. A series of investigations revealed that illegal CFC-11 production and use originated from Eastern China.

While the Chinese government investigated and promptly arrested a group of industrialists violating the Montreal Protocol, this unfortunate situation is neither the first nor the last time a treaty violation occurred. In this case, the offense prevents the ozone hole from closing and increases the amount of greenhouse gases in the upper atmosphere. Equally important, it reminds us that merely creating treaties does not end the threat to our environment. States have essential roles in domestic implementation; likewise, corporations, NGOs, and scientists directly impact the quality of implementation and, therefore, environmental protection.

References

1. Annan KA (2000) We the peoples: the role of the United Nations in the 21st century. United Nations, New York
2. Andersen SO, Sarma KM (2002). Protecting the ozone layer: the United Nations history. Earthscan, Hoboken
3. Benedick RE (1998) Ozone diplomacy. Harvard University Press, Cambridge
4. Cook E (1990) Global environmental advocacy: citizen activism in protecting the ozone layer. Ambio 19(6/7):334–338
5. Doolittle DM (1989) Underestimating ozone depletion: the meandering road to the Montreal Protocol and beyond. Ecology LQ 16(2):407–441

6. Elliott L (2016) Smuggling networks and the black market in ozone depleting substances. In: Wyatt T (ed) Hazardous waste and pollution. Springer, Heidelburg, pp 45–60
7. EPA (1978) Government ban on fluorocarbon gases in aerosol products begins October 15. https://archive.epa.gov/epa/aboutepa/government-ban-fluorocarbon-gases-aerosol-products-begins-october-15-1978.html. Accessed 14 Dec 2021
8. Epstein G et al (2014) Governing the invisible commons: ozone regulation and the Montreal Protocol. Int J Commons 8(2):337–360
9. European Environmental Agency (2020) Maximum ozone hole extent over the southern hemisphere from 1979 – 2019. https://www.eea.europa.eu/data-and-maps/figures/maximum-ozone-hole-area-in-7. Accessed Jan 8 2022
10. Farman J et al (1985) Large losses of total ozone in Antarctica reveal seasonal ClOx/NOx interaction. Nature 315:207–210
11. Green BA (2009) Lessons from the Montreal Protocol: guidance for the next international climate change agreement. Environ Law 39(1):253–283
12. Haas PM (1990) Saving the Mediterranean: the politics of international environmental cooperation. Columbia University Press, New York
13. Hofman J, Gibbs M (1988) Future concentration of stratospheric chlorine and bromine, EPA, Set. No. 400/1-88/005
14. Implications of the findings of the expedition to investigate the ozone hole over the Antarctic, 100th Congress.(1987) (testimony of F. Sherwood Rowland)
15. Lemonick MD (1987, October 19) The heat is on: chemical wastes spewed into the air threaten the earth's climate. Time Magazine. http://content.time.com/time/subscriber/article/0,33009,965776-5,00.html. Accessed 14 Dec 2021
16. Litfin K (1994) Ozone discourses: science and politics in global environmental cooperation. Columbia University Press, New York
17. Makhijani A, Gurney KR (1995) Mending the ozone hole science, technology, and policy. MIT Press, Cambridge
18. McGrath M (2019, May 22) Banned CFCs traced to China say scientists. https://www.bbc.com/news/science-environment-48353341. Accessed Jan 8 2022
19. Molina M, Rowland F (1974) Stratospheric sink for chlorofluoromethanes: chlorine atom-catalysed destruction of ozone. Nature 249(5460):810–812
20. Moore CA (1990) Industry responses to the Montreal Protocol. Ambio 19(6/7):320–323
21. Multilateral Fund (2022) Multilateral fund for the implementation of the Montreal Protocol. http://www.multilateralfund.org/default.aspx. Accessed Jan 8 2022
22. NASA (2020) NASA Ozone watch images, data, and information for atmospheric ozone. https://ozonewatch.gsfc.nasa.gov/. Accessed 8 Jan 2022
23. New York Times (1987, September 29). Washington talk: briefing; 'styro-wars'. Section A, p 30
24. Parson EA (2003) Protecting the ozone layer: science and strategy. Oxford University Press, New York
25. Patchell J, Hayter R (2013) How big business can save the climate: multinational corporations can succeed where governments have failed. Foreign Aff 92(5):17–23
26. Sand PH (1985) Protecting the ozone layer: the Vienna Convention is adopted. Environ 27(5):18–43
27. Spurgeon D (1997) Ozone treaty 'must tackle CFC smuggling.' Nature 389(6648):219–219
28. UNEP (1981a) Report of the Governing Council of the United Nations Environment Programme on the work of its ninth session. A/36/25
29. UNEP (1981b) Montevideo programme for the development and periodic review of environmental law. No. 88-5614-0281E
30. UNEP (1987) Ad hoc working group of legal and technical experts for the preparation of a protocol on chlorofluorocarbons to the Vienna Convention for the Protection of the Ozone Layer (Vienna Group). UNEP/WG.167.2
31. UNEP (2011) Risk assessment of illegal trade in HCFCs. http://www.pnuma.org/ozono/publicaciones/HCFC%20illegal%20trade%20report%20web%20version.pdf. Accessed Jan 8 2022
32. Vienna Convention, 22 Mar 1985. https://treaties.un.org/doc/Treaties/1988/09/19880922%2003-14%20AM/Ch_XXVII_02p.pdf

Chapter 6
Regulating the Movement of Hazardous Waste

Abstract Industrial manufacturing processes create hazardous chemicals not only as a finished product, but also as a hazardous waste. Disposing of these wastes occurs globally, resulting in human health and environmental impacts. Beginning with the Love Canal tragedy in the United States, this chapter describes the hazardous waste regime. As United States domestic laws restricted disposal techniques, producers sought disposal locations in other countries, creating a toxic trade in hazardous waste. Consequently, international environmental diplomats created the Basel Convention on the Control of Transboundary Movements of Hazardous Waste and Their Disposal that mandated importing countries' consent before transportation of hazardous waste to a foreign shore. Additionally, the text provides a brief overview of the environmental justice movement that began when Dr. Robert Bullard realized minority communities were more highly exposed to toxic chemicals, resulting in higher levels of environmental disorganization in these communities. Further, this pattern replicated internationally along the lines of the North–South gap. The chapter concludes by providing a synopsis of Robert Putnam's two-level game as an explanation for domestic sources of international treaties.

Keywords Hazardous waste · Domestic policy · Basel Convention on the Control of Transboundary Movements of Hazardous Wastes and Their Disposal · Environmental justice · Love Canal · Toxic trade

This chapter examines a series of United States domestic environmental laws followed by a discussion of the most infamous environmental disaster involving hazardous waste disposal, the Love Canal. This incident started an international movement that began examining the relationship between environmental health, hazardous waste disposal, and the transboundary movements of hazardous waste.

In the process of shipping this toxic trade around the world, geospatial patterns of environmental risks, human health impacts, and degraded environments gave rise to conceptualizations of environmental justice. The need for environmental justice for minority communities, not just in the United States but around the world, stemmed from the willingness of the corporate and government elites to take advantage of the politically disempowered and financially impoverished communities.

In order to understand the international negotiating history surrounding the trade in hazardous waste, The Basel Convention must first be explored. The first section highlights the hazardous waste problem, including providing details about the United States Resource Conservation and Recovery Act (RCRA) and the TSCA, two pieces of domestic legislation that heavily influenced negotiators examining this issue. Section two provides an overview of the Basel Convention. Section three looks at later amendments such as the Ban Amendment and the Basel Protocol on Liability and Compensation for Damage Resulting from Transboundary Movements of Hazardous Wastes and their Disposal (Basel Protocol). Section four examines international environmental justice, an issue closely associated with the transboundary movement of hazardous chemicals, including waste. Section five concludes the chapter by examining the interaction between national and international politics through the academic concept of a two-level game.

6.1 Hazardous Waste Laws Within the United States

In October 1976, Republican President Gerald Ford signed two essential pieces of Congressional legislation regulating hazardous chemicals within the United States; TSCA on October 11 and RCRA on October 21. Congress drafted these two laws in response to the numerous environmental health incidents that stemmed from close contact with chemicals in the workplace. Of the two laws, RCRA created the more stringent regulatory framework.

TSCA granted the EPA regulating authority over chemical substances that were not regulated under other federal laws, allowing the EPA to stop production and force a withdrawal of a chemical substance from the national market. TSCA also required manufacturers to give EPA notice through a pre-manufacturing notice 90 days before the beginning of manufacture or importation into the United States. Chemical manufacturers and importers must share information about the environmental health impacts of these chemicals to determine if unreasonable risk or injury could occur due to chemical exposure.

TSCA established a notification system for the transboundary movement of chemicals, including imports and exports. Upon receipt of a TSCA export notification, the EPA notifies other countries when companies intend to export a chemical's first shipment each year to a country. Under United States domestic law, companies may manufacture chemicals for use in other countries that have been banned domestically. Similarly, entities wishing to import chemicals into the United States must also notify the EPA. All incoming chemicals must also meet all domestic regulatory requirements.

RCRA amended the Solid Waste Recovery Act of 1965, the first law within the United States to improve solid waste disposal habits. Congress passed the RCRA upon realizing the extent of chemical contamination across the United States. RCRA forces waste producers to manage hazardous chemicals in a manner that protects human health and the environment, including by reducing or eliminating waste production. Congress also intended to encourage recycling to conserve energy production and raw materials [13].

RCRA required cradle-to-grave management of hazardous materials within the United States. Chemicals triggered RCRA regulations by appearing on the named chemical list within the statute or exhibiting one or more hazardous waste characteristics of ignitability, corrosivity, reactivity, or toxicity. RCRA utilizes an intense tracking system where EPA assigns a code to each waste stream from the moment of generation to disposal. This code must be included on a manifest that identifies the waste and the amount of waste contained in the shipment. This paperwork originates with the generator and specifies all transporters. The manifest accompanies the waste from the generator to the final treatment, storage, and disposal facility.

The passage of RCRA encouraged companies to be more careful when storing, transporting, and disposing of their waste streams to avoid contamination in the future. RCRA represented a significant tightening of the regulation of hazardous waste within the United States that escalated the price of disposing of hazardous waste. Estimates contemporary to the time espouse that disposal costs changed from $2.50 to bury a ton of waste before RCRA to $200 after RCRA passed. Similarly, prices to burn a ton of waste increased from $50 to $2000 [32]. RCRA, did not, however, force polluters to clean up already contaminated waste sites.

In 1978, national television introduced the country to the Love Canal subdivision. The subdivision was located just four miles south of Niagara Falls, New York. The Love Canal neighborhood, once the site of the American dream, turned into an American nightmare. Unsuspecting homeowners bought into this subdivision on land sold by the Niagara Falls School Board expecting to enjoy their home while their children attended the 99th Street Elementary School.

Love Canal began as an industrial project in the late nineteenth century. Community developers planned to divert water from the Niagara River to form a new canal as a central feature in a new community. A hydroelectric power station would use the diverted water to produce electricity. Construction began on the canal, but an economic downturn ended the project. The Love Canal turned from an abandoned industrial infrastructure project into a municipal landfill.

Gibbs [10] reported that the city and the United States Army utilized the site. Additionally, the Hooker Chemical Company primarily utilized the site to dispose of chemicals used to manufacture a variety of chlorinated hydrocarbons, dyes, and caustic products. The company loaded the chemical waste into 55-barrel drums and deposited the drums into the canal. Hooker Chemical Company eventually bought the canal turned landfill in 1947. Hooker utilized the site for roughly ten years before closing the dump by installing a clay liner over the top.

In 1953 Hooker sold the property, now filled with toxic chemicals, for $1 to the Niagara Falls School Board for a new elementary school that opened in 1955. The school sat on the edge of a highly potent and dangerous toxic chemical dump. Worse still, the buried chemicals were directly underneath the playground. Unsuspecting home buyers saw the school's proximity to the newly built homes as an advantage of living in the Love Canal neighborhood. Instead, the playground and school became emblematic of the problems stemming from hazardous waste disposal.

By 1958, the chemicals stored in the drums escaped containment. The chemical leaks became apparent on the surface of the various properties close to the school, and reports of children receiving chemical burns began [22]. Over the next twenty years, a steady stream of complaints about chemicals became normalized for this community. Additionally, the health of children and adults near the landfill declined.

New York's State Department of Environmental Conservation began an investigation of the numerous complaints involving this neighborhood in 1977. In 1978, local journalist Mike Brown authored multiple newspaper articles about the contamination in the Niagara Falls Gazette. After hearing the news reports, local residents, including Lois Gibbs, became concerned. Gibbs emerged as the neighborhood leader seeking answers from state officials about the environmental health impacts of living near a toxic waste dump after confronting state officials about the unusually high number of congenital disabilities, illnesses, and suspicious maladies located in a cluster around the canal. With Gibbs as a vocal leader, the Love Canal Homeowners Association advocated for a complete buyout of the homes in the neighborhood as the property values collapsed due to the contaminated land and the publicity generated by the protests.

By August 1978, Love Canal became a national issue, and then President Jimmy Carter issued an emergency order to try to clean up the contamination. A few days later, the state of New York, acting in conjunction with the federal government, began relocating families closest to the school out of the area. The relocation process included the reimbursement for the value of the home.

The Love Canal story shocked the nation and reverberated around the globe. While other sites such as the Valley of the Drums in Kentucky contained more pollution, Love Canal became the most infamous. The global public realized that environmental health risks due to exposure to toxic chemicals occurred regularly. Lois Gibbs became a national heroine and leader within the environmental health movement. She founded the Center for Environment, Health, and Justice to continue advocating on behalf of individuals and communities negatively impacted due to the proximity of hazardous chemicals.

The public outrage caused by Love Canal spurred major United States legislation as Congress passed the Comprehensive Environmental Response, Compensation and Liability Act (CERCLA), better known by its nickname, Superfund. CERCLA would force companies to pay to clean up areas they had contaminated, even if the company acted in compliance with disposal laws at the time. Superfund monies come from various sources, including state budgets, federal taxes, and cost recovery from potentially responsible sources. Unsurprisingly, Love Canal became the first location on the National Priorities List for a Superfund cleanup.

Love Canal also raised awareness of the health problems that originated from chemical exposure. Not in my backyard (NIMBY) became associated with resistance to undesirable land uses, including hazardous waste, chemical facilities, and sewage outfalls. As grassroots organizations spoke with each other, they recognized that there were, in fact, many, many backyards [12]. Thus, a new acronym became associated with hazardous waste management, and more specifically, hazardous waste siting. Not in anybody's backyard (NIABY) reflected the realization that the best protection against environmental health problems stemming from hazardous waste disposal involved reducing the hazardous waste, with zero hazardous waste as the ideal. In other words, the best protection from hazardous waste is to quit producing the waste.

As the environmental health movement successfully blocked the creation of new facilities or forced the closure of existing facilities, the hazardous waste nevertheless required a disposal method. The changes in the United States waste disposal market caused by the combination of TSCA, RCRA, and CERCLA forced some facilities to close down, contributing to the rise in prices for disposing of hazardous waste.

Consequently, producers began looking at alternative disposal mechanisms both domestically and internationally. Corporations sought to take advantage of differences in national laws that allowed corporations to lower their disposal costs. This phenomenon may be described as the race to the bottom, where countries lower the cost of conducting business as a means of attracting wealth to their jurisdiction. For a developing country with little means of generating wealth, the financial payments represented an opportunity to pay back loans or create much-needed infrastructures like hospitals, schools, and roads.

The so-called "toxic trade" in hazardous chemicals exposed the developing countries to chemical wastes without knowing the chemical composition or health impacts. Additionally, illegal dumping also occurred regularly, meaning that corporations shipped the waste to developing countries without their consent. The corporations dumped the material regardless. Additionally, bribing a corrupt official or disguising wastes as useful materials also occurred as part of this trade.

Strohm [31] proclaimed 1980 as the beginning of the international waste trade. She points out that no economic analysis for this industry exists due to differences in the definition of hazardous waste, making this claim difficult to prove or disprove. However, the Organization for Economic Co-operation and Development (OECD) attempted to fill this knowledge gap by estimating the amount of hazardous waste disposed of as a precursor to projecting the transboundary hazardous waste trade. Clapp [8] pointed out that economic systems and environmental quality are intrinsically linked, with the hazardous waste trade providing one specific example of this linkage.

Scholars likely accepted an OECD estimate for the transboundary movement of hazardous waste, indicating that roughly 5.2 million tons left the OECD countries from 1986 to 1990, destined for Eastern Europe and developing countries [16, 40]. Montgomery [20] believed this estimate is too high and insists that less than 1% of the hazardous waste by volume left the United States when states convened to negotiate the Basel Convention. However, given the overall amount of hazardous waste produced by the United States and the hazardous waste generated by other OECD countries, this amount of hazardous waste could significantly jeopardize the environmental health of unprepared recipients. Additionally, the toxicity of the hazardous waste also remained unknown and likely varied according to the individual shipment.

While trades could consist of North–North trades and South-South trades, the most concerning trade pattern involved North–South trades. These shipments leave the developed countries with more stringent regulations, bound for countries with less strict rules [19]. The Southern countries had and continue to have widely varying domestic laws regulating these wastes' transport, storage, and disposal. Further, these countries may not possess the same resources to control, eliminate, and remediate the eventual environmental damage that inevitably arises from accidental releases, poor disposal practices, and illegal dumping.

6.2   Toward the Basel Convention

The academic literature highlighted and continues to highlight situations that resulted in environmental damage caused by the hazardous waste trade [8, 21, 24]. Perhaps the most infamous of these situations involved the garbage barge known as the Khian Sea. The Khian Sea incident involved elements of fraudulent behavior. The owners of the Khian Sea effectively ignored the sovereignty of Haiti and violated international environmental norms and principles.

The Khian Sea accepted a subcontract to dispose of municipal incinerator ash from Philadelphia, Pennsylvania, United States. Shippers originally arranged for the material to enter the Bahamas as fill material in 1986. However, toxicity concerns about the municipal incinerator ash emerged, and the Bahama government refused its permission to dump the ash.

The Khian Sea set out on a two-year search for a port to accept the waste but failed as multiple Caribbean states refused to accept the material. The ship eventually dumped part of the ash illegally in Gonaives, Haiti, in January 1988. The Haitian government attempted to force the Khian Sea to take the ash back onboard, but the ship left the harbor.

The Haitian government did not have the resources to remove the ash, so it sat on the beach. Chemical testing revealed the presence of heavy metals such as lead and cadmium, along with dioxin, a known carcinogen [28]. These metals leached into the environment before a containment facility was built to hold the waste [4].

Accounts of the Khian Sea's itinerary vary slightly, but the ship appears to have visited five continents and changed its name twice before sailing into Singapore as the Pelicano. However, the ash stored onboard the vessel had disappeared. The owners of the Khian Sea, William P. Reilly and John Patrick Dowd stood trial in the United States for perjury in 1993. During the trial, the garbage barge captain, Arturo Fuentes, admitted to dumping the remainder of the ash in the Atlantic and Indian oceans.

After staying in Haiti for almost ten years, the waste from the Khian Sea returned to the United States. Years later, Louis Paolino, whose company Joseph Paolino and Sons underwent a federal investigation as part of the Khian Sea incident, moved to Eastern Environmental Services, a garbage hauling company. Eastern Environmental Services bid on a contract in New York City. As part of the contract, New York City insisted on reimporting the waste from Haiti to the United States. Eastern Environmental Services agreed to pay part of the return trip costs. Environmental groups launched a campaign, "Return to Sender," that successfully returned the waste to a landfill outside of Philadelphia for final disposal [24].

While the Khian Sea undoubtedly created public support for the movement to regulate transboundary hazardous waste, the idea for a convention originated much earlier. This effort stemmed from the Montevideo Environmental Law Program. UNEP commissioned an expert group to review topics UNEP could promote in creating new environmental law either through treaty or encouraging countries to develop their domestic law [34]. The Senior Government Officials Expert in Environmental Law met in Montevideo, Uruguay, from October 28 to November 6, 1981, to finalize an agenda for UNEP that effectively served as a long-term strategic plan for the organization.

UNEP's Governing Council accepted this recommendation and created an Ad Hoc Working Group of Experts on the Environmentally Sound Management of Hazardous Wastes and charged the group with developing guidelines for further consideration. This Working Group produced a report for the UNEP's Governing Council review in its 1987 session. Known as the Cairo Guidelines and Principles for the Environmentally Sound Management of Hazardous Wastes, UNEP's Governing Council adopted Decision 14/30 on June 17, 1987.

Diplomats from 24 states organized themselves for the upcoming negotiations through an organizational meeting in Budapest, Hungary, in October 1987. This session produced the First Draft Convention on the Control of Transboundary Movements of Hazardous Waste using the Cairo Guidelines and Principles for the Environmentally Sound Management of Hazardous Wastes as the starting point [26].

Early issues identified within the treaty's scope included the fact that some states did not have a legal definition of hazardous waste, much less a legal framework for managing the waste. Additionally, states recognized that other international treaties and bilateral and regional agreements might already exist that included regulating specific types of hazardous waste. Diplomats also identified the type of liability and how a penalty should be imposed as a complex topic that may not be resolved in time for inclusion in the initial treaty text [35].

Diplomats and observers met in five sessions in 1988 and 1999. Session one occurred from February 1–5, 1988, in Geneva, Switzerland. Additional meetings occurred from June 6–10, 1988, in Caracas, Venezuela, from November 7–16 in Geneva, Switzerland, from January 30–February 3, 1989, in Luxembourg, and from March 13–17, 1989, in Basel, Switzerland. As is typical with a negotiating committee, fewer states participated at the beginning of the working group than at the end. Thus, state participants grew from 33 countries in Geneva in 1988, to 96 countries attending one or more of the Working Group sessions. NGO attendance, however, remained low and inconsistent throughout the duration of the Working Group. Only 23 NGOs attended one or more Working Group meetings. For example, the Chemical Manufacturers Association participated in the Working Group negotiating sessions, but not the organizational session.

Initially, attendees intended to finish negotiations by the end of the third session [36]. However, deep divisions between African and industrialized countries nearly derailed negotiations. During the negotiations, states disagreed on the need to negotiate for a complete ban or a partial ban. The United States and its allies argued for a partial ban, as they did not want to hinder trade in recovery or recycling processes.

In contrast, the African countries, through the Organization of African Unity, the precursor to the African Union, argued for a total ban on the trade of hazardous waste. The Organization of African Unity meeting in May 1988 condemned the practice of disposing of hazardous waste on the continent and called for widespread participation by the African countries to bolster the position of states calling for a complete ban on hazardous waste trade. Kummer [17] reported that the division between the two negotiating groups only reconciled upon the personal efforts of UNEP's Executive Director Mostafa Tolba, who initiated an informal Working Group meeting to resolve these differences and allow conference negotiations to continue.

The Working Group completed its meetings and forwarded draft text, albeit with bracketed text indicating areas of disagreement for further consideration to the next meeting. The Basel Convention negotiating session lasted a mere three days, an extremely short time frame for concluding a diplomatic conference. Diplomatic differences dominated the discussion like the Working Group meetings that preceded this meeting. The Organization of African Unity, in its opening statement, declared that anything less than a complete total ban on the movement of hazardous wastes would not be acceptable, effectively jeopardizing the Convention outcome [17].

The Basel Convention regulates the movement of hazardous waste, solid waste, and municipal incinerator ash by detailing the circumstances under which this trade is permissible. Under the treaty's terms, hazardous waste means all wastes listed in Annex I of the Convention and any wastes that meet the characteristics of hazardous waste as defined in Annex III. Additionally, any wastes specified as hazardous waste under the domestic law of a party that engaged in the export or import of the hazardous waste, the country of import, or any states through which the waste travelled on the way to its final disposal destination. States also defined household waste and municipal incinerator ash from burning household waste as waste regulated by the Convention in Annex II.

This treaty does not cover radioactive wastes and wastes from a ship operating under normal conditions because these items are regulated by the IAEA and the MARPOL, respectively. If the chemical is not a waste, the Basel Convention may not cover its transport. Chemicals that are destined for use in another country, even if that chemical is banned in its country of origin, may still be sold to another country.

Given that the developed countries tend to have more stringent regulations regarding hazardous waste, the Basel Convention's definition applies as the party of export. In the United States, the RCRA defines hazardous waste. However, RCRA utilizes a loophole around recycling that is not present within the Basel Convention. Under RCRA, materials destined for recycling are not considered hazardous waste, even if the physical and chemical properties meet the definitional qualities of hazardous waste. This recycling exemption encouraged sham recycling, where corporations gave an impression of sending the waste off to be recycled but instead disposed of it. The Basel Convention handles this situation differently by including hazardous waste recycling in the Basel Convention [37]. This inclusion could be used to limit shipping materials overseas for recycling processes, effectively ending the practice of recovering valuable materials for future uses [27].

Under the Convention, states agreed to end shipments to all countries that decline to accept hazardous waste. Two forms of action can be utilized. Countries may permanently ban the importation of hazardous waste, or states may reject a specific shipment of hazardous waste. The Basel Convention prohibits trade to any party that is not a party to the Basel Convention. The Convention bans the movement of hazardous waste from a Basel Convention signatory to a Basel Convention non-signatory state unless a treaty is in place that is at least as stringent as the Basel Convention. Shipping hazardous waste across national boundaries revolves around the prior informed consent (PIC) mechanism where the shipping country notifies the receiving country in writing about the pertinent details of the shipment, including the nature of the waste, its disposal method, transportation details, and the relationship with the disposer. Receiving states must give their written consent before the shipment begins its trip to the receiving country. Parties may not legally ship hazardous waste to a developing country without receiving PIC.

The Basel Convention also requires states to detail the characteristics of the chemicals hazardous waste imports and exports as part of the regulatory scheme. Thus, the Basel Convention requires states participating in the hazardous waste trade to create and implement a mechanism for tracking their hazardous waste, similar to the RCRA manifest scheme. If an international shipment cannot be completed, the Basel Convention requires the waste to return to its country of origin or ensure its safe disposal in another locale. In many countries, this function requires legislation to provide funding for the personnel to carry out this function.

The Basel Convention requires that states manage their waste in an environmentally sound manner. Environmentally sound manner, as defined in the Basel Convention, "means taking all practicable steps to ensure that hazardous wastes and other wastes are managed in a manner which will protect human health and the environment against the adverse effects which may result from such wastes" [2]. This admittedly vague wording meant that states should manage their waste in such a way as to be protective of human health as well as environmentally benign. In practice, this term may mean a different level of protection based on the vulnerability of individuals and communities.

The Basel Convention also encourages states to minimize the amount of hazardous waste moving around the globe. This minimalization effort could take two forms. First, countries should lower the amount of hazardous waste it produces. Second, countries should dispose of hazardous waste as close as possible to the locale that created the waste.

The establishment of international law that regulates the transboundary movement of hazardous waste also gave rise to the illegal disposal of hazardous waste. Illegal could be as simple as a failure to notify appropriate authorities about shipment, or it could be more nefarious if the waste was dumped without permission in another country. Thus, UNEP works with various international organizations to limit international crime, such as the International Criminal Police Organization (INTERPOL).

Assigning responsibility for illegal actions carried out by a national created sharp disagreement among the various groups during the convention [17]. Exporting states may be held responsible for nationals that violate terms and conditions relating to generation and exportation, while importing states may be held responsible for nationals that violate terms and conditions relating to importation and disposal.

The Secretariat of the Basel Convention resides in Geneva, Switzerland. States assigned various roles to the Secretariat, including giving technical advice, convening meetings on behalf of the COP, and connecting parties to the Convention to appropriate resources, including resources for capacity building and technology transfers. Additionally, UNEP may assist in identifying illegal waste trafficking but UNEP does not have any ability to halt the illegal waste trade. Further, UNEP does not act as a regulatory agency in that it does not inspect, audit, or otherwise verify the accuracy of transboundary shipments.

Parties under the Convention agreed to establish regional training and technology centers located in various regions around the globe. The fourteen centers assist states, particularly developing countries, with meeting their obligations under the convention by providing training and other relevant information about toxic chemicals, hazardous waste disposal, pollution control technologies, and best practices relating to the Basel Convention and other later treaties that collectively regulate toxic chemicals internationally.

The Basel Convention utilized a low threshold for ratification, with a mere twenty countries required to agree to the framework [2]. This low number stands in stark contrast to the 96 countries that worked on the draft treaty or the 193 member states within the UN. More tellingly, none of the developed countries traditionally responsible for shipping waste overseas needed to ratify the treaty before it entered into force on May 5, 1992. Thus, one of the minor issues at the Earth Summit in Rio de Janeiro, discussed in Chap. 7, included encouraging more states to ratify the Basel Convention.

Various states hesitated to sign the provisions of the Basel Convention. African states viewed the Basel Convention as allowing Northern countries to continue poor environmental disposal practices that unfairly jeopardized human health on the African continent. Other developing states agreed with this position. However, Asian countries have been more willing to accept hazardous waste trade than African countries [3].

States that exported hazardous waste also did not favor the treaty initially. For example, the United States is not currently a party to the Basel Convention. While the United States Senate responsible for advising and consenting to international treaties did so, the president may not provide formal written notification to the UN until Congress passes the legislation needed to comply with the treaty. The rationale for the refusal to ratify is unique. The United States Senate ratified the treaty but did not submit the paperwork to the UN because the United States would be out of compliance with the Basel Convention.

Schmidt [27] posited that RCRA would need to be amended to create the necessary compliance mechanisms. The structure of the Basel Convention also potentially expands EPA's role in international relations beyond its current mandate [3]. Schmidt [27] further notes that RCRA and the Basel Convention do not align well, particularly in areas regarding the definition of hazardous waste and recycling.

Prominent NGOs such as Greenpeace and the Basel Action Network both view the creation of the Basel Convention as legitimizing the transboundary hazardous waste trade [21, 27]. Clapp [8] pointed out that prominent environmental groups formed a coalition with developing country states, which she named the Third World—NGO alliance. This network combined to support a total ban on hazardous trade. NGOs offered their technical expertise to developing countries that promoted the NGOs preferred policy outcomes as part of the treaty text. This symbiotic relationship required that developing states keep their NGO advisers informed of the content of the supposedly closed-door negotiations.

As states typically do not support strong environmental enforcement activities at the international level, voluntary mechanisms such as those provided by NGOs like Greenpeace, succeed in highlighting areas of non-compliance that deter others from repeating the same non-compliant activities [7]. NGOs' contacts with the national media improved the likelihood of compliance with hazardous waste rules and regulations both domestically and internationally.

Scholars critiqued the Basel Convention for its failure to focus on waste prevention [21] and for de facto approval of the hazardous waste trade via creating a regulatory scheme to track its movement [17, 26]. Corporations have little incentive to redesign their processes to eliminate the creation of hazardous waste when disposal costs are low. Conversely, corporations that face high disposal costs may seek to lower these costs by implementing manufacturing process changes, including raw materials substitutions.

Krueger [16] pointed out that the Secretariat recommended standardized paperwork for the PIC process, but the treaty did not require states to follow this recommendation. Further, he noted that African countries' regulatory staff did not possess the appropriate technical competencies necessary to make rational choices regarding environmental risk management.

Perhaps the most contentious issue, liability for improperly disposed of wastes, remained unresolved after diplomats finalized the text for the Basel Convention [7, 21]. Okaru [21] stated that the absence of a liability scheme severely weakened the Basel Convention. Developing countries adopted a negotiating stance that asked for a strict liability standard. The developed country of origin took responsibility for all damages that occurred in the movement and disposal of the waste. Strict liability means that the defendant is legally responsible for damages, regardless of the defendant's intent to cause harm or injury.

6.3 New Developments

After completing the negotiations in Basel, Switzerland, many states were unsatisfied with the treaty text [7]. African states felt that the removal of a complete ban on the hazardous waste trade left their states exposed to both legal trade and illegal dumping. Given their inability to finance or conduct remediation operations, the African people believed they would be exposed to higher levels of environmental risk than their counterparts worldwide. At the other end of the spectrum, developed countries shared concerns about the impact of the Basel Convention on their domestic companies [21]. These concerns included price increases for hazardous waste disposal and worries that the Basel Convention would discourage recycling waste materials. Additionally, the Organization of African Unity expressed concern about the ease of circumventing the Basel Convention due to the poor administrative skills in many developing countries [30].

Twelve African states, utilizing the cooperative mechanisms provided by the Organization of African Unity, met together to write their own treaty regarding the hazardous waste trade, culminating in the Bamako Convention on the Ban of the Import into Africa and the Control of Transboundary Movement and Management of Hazardous Wastes within Africa (Bamako Convention) in Bamako, Mali in January 1991. This treaty operates in addition to the Basel Convention as a regional agreement as specified in Article 11 of the Basel Convention. This treaty effectively strengthened environmental protection for its signatories by banning imports of hazardous waste, including for recycling, and limits hazardous waste trading between African countries.

The Bamako Convention negotiating process proved to be less contentious than the Basel Convention due to the similar circumstances of the negotiating parties. In an intriguing twist for international affairs, state sovereignty worked in favor of environmental protection when Southern states exerted their sovereignty to end the hazardous waste trade to their homelands. African nations gathered to create the Bamako Convention that banned the entry of hazardous and radioactive wastes into the African continent. Additionally, the Bamako Convention improved the sophistication of the regulatory scheme to deter non-compliance and instituted stricter liability requirements for those participating in the hazardous waste trade.

Under the Bamako Convention, African states removed the exemptions for specific types of hazardous waste regulated by other conventions, such as radioactive waste. These states also stipulated that wastes that contained a known hazardous waste or exhibited a characteristic of hazardous waste would be covered by this regional treaty. African states also removed the territorial exclusion waters under national jurisdiction. Kreuger [16] pointed out that the Bamako Convention altered PIC paperwork processes by requiring African countries to submit all paperwork to the Basel Secretariat. In doing so, the Bamako Convention created transparency and allowed for oversight by a secretary with greater administrative capabilities.

African states also settled the question of legal liability stemming from the hazardous waste trade. The Bamako Convention established both unlimited liability and joint and several liability [14, 30]. Unlimited liability established states' ability to recover all losses related to removing the offending wastes, remediation of the environment, recovery of legal costs, and a punitive fine to discourage future infractions. This agreement occurred, in part, because African countries tend to receive waste rather than create waste. The similar circumstances allowed negotiators to agree more rapidly as circumstances made it unlikely any African country would be subject to the liability procedures.

The Bamako Convention entered into force on April 22, 1998. This document serves as a representative of regional treaties that ban hazardous waste trade into global regionals. It is not, however, the only regional treaty of this type. Other regional treaties that ban hazardous waste include the Convention to Ban the Import into Forum Countries of Hazardous Waste and to Control the Transboundary Movement and Management of Hazardous Wastes within the Pacific for the South Pacific Region, and Article 39 of the Lomé IV Convention.

Krueger [16] analyzed hazardous waste trade data in the period after the implementation of the Basel Convention. He found that shipments in wastes for disposal declined while shipments of wastes intended for recycling rose. Kellenberg and Levinson [15] tracked the hazardous waste trade from 1988 to 2008, including the period before and after negotiations of the Basel Convention and the Ban Amendment. Their research concluded that absolute amounts of hazardous waste continued to move between countries through 2008. The authors, however, hesitate to state that the agreements did not make an impact on the hazardous waste trade as some evidence suggests that OECD countries that ratified the Ban Amendment opted to transport less toxic waste.

One mechanism that states could utilize to augment the strength of the Basel Convention included introducing liability principles that assigned financial responsibilities to parties that damaged the environment. Resolution 3 from the Conference of Plenipotentiaries directed the UNEP Executive Director to create an ad hoc working group to draft text for states to consider at the first meeting as the COP.

States also expressed a clear preference for continuing conversations about banning transboundary hazardous waste. COP 1 convened after the Earth Summit, a twenty-year celebration of the UNCHE conference, in Piríapolis, Uruguay, from December 2–4, 1992. Clapp [8] credited Greenpeace with keeping alive the hope of banning the hazardous waste trade at this meeting as states agreed to consider the issue in the next three years. The decision to create the ban on the movement of hazardous waste, including recycling, from OECD countries to non-OECD countries occurred during COP 2 in Geneva, Switzerland, on March 21–25, 1994.

The third meeting of the COP, from September 18–22, 1995, in Geneva, Switzerland, formalized the Ban Amendment in 1995. The Ban Amendment prohibits the exportation of all hazardous waste from the EU, OECD countries, and Lichtenstein to non-OECD countries. Notably, the Ban Amendment does not limit the exportation of hazardous waste between these countries. Thus, a shipment of chemical hazardous waste from the EU to the United States for disposal would be allowed under the Ban Amendment. The Ban Amendment entered into force in December 2019. Notably, the United States is not a party to the Ban Amendment.

Four years later, in December 1999, states finalized the Basel Protocol that determined liability at COP 5 in Basel, Switzerland. Choksi [7] reviewed the prospects that states would ratify the Basel Protocol and concluded that the document created a highly flawed method of assigning liability. The many loopholes within the document create opportunities for waste generators to avoid liability in its entirety as the Protocol assigns strict liability to notifiers rather than generators before the disposer takes possession of the waste. After the disposer takes possession of the waste, the disposer becomes strictly liable for any harm. Damages that could be actionable under a strict liability claim include personal injury, damage to personal property, loss of economic income, remediation costs, and preventative measures. Parties under the Convention that trade with non-Parties may be subject to fault liability without a maximum cap for the state that signed the Basel Convention.

States also attempted to limit their financial exposure by capping the amount payable for civil penalties by declaring a national liability cap for strict liability claims only. In order to avoid states setting an arbitrarily low maximum claim amount, the Basel Protocol established a minimum claim amount that is proportional to the hazardous waste traded. Additionally, organizations participating in the hazardous waste trade must carry insurance to fund payments fully, if needed.

As of the end of 2021, twelve states, all developing countries, completed ratification falling well short of the twenty states needed to complete ratification. Perhaps more telling, no developed country has ratified the Basel Protocol. However, environmental groups, the Basel Convention Secretariat, and other interested parties periodically encourage states to ratify the Basel Protocol.

After creating the Basel Protocol, states turned to implementing the Basel Convention. Parties to the Convention established Basel Convention Regional Centers to assist developing countries with specific questions about hazardous waste transport and management. A critical element of this implementation included the creation of technical guidelines for the environmentally sound management for specific categories of waste, including, but not limited to, battery wastes, plastics, dismantling of ships, and e-waste.

6.4   International Environmental Justice

The toxic trade brings to light the extreme difficulties in exposure to environmental risks and the uneven environmental burdens imposed by the toxic trade. Environmental justice captures the idea that the global South suffers from environmental burdens that it did not cause, cannot control, and cannot avoid. Conceptually, environmental justice emerged as a social issue within the United States in the 1980s and 1990s as citizens and scholars correlated the presence of landfill disposal facilities in African American communities.

In 1982, a protest in Warren County, North Carolina, broke out over a decision to site a landfill in an African American Community. While the protest failed to stop the facility from hosting polychlorinated biphenyl (PCB) contaminated soil and other toxic chemicals, the protests marked the beginning of the environmental justice movement. After the Warren County protest, further evidence of environmental discrimination quickly emerged. Bullard [5] released a study of municipal landfill disposal patterns in Houston that correlated the siting of hazardous waste facilities in African American neighborhoods.

One of the protestors arrested during this demonstration, the Reverend Benjamin Chavis Junior, encouraged the United Churches of Christ to study the locations of other hazardous waste landfills. This report, released as the Toxic Wastes and Race in the United States in 1987, provided shocking documentation that correlated the placement of hazardous waste landfills in African American communities [38]. Demographic characteristics of communities more likely to suffer from disproportionate environmental risks and human health impacts may be characterized as ethnic minorities, communities experiencing low income and/or high poverty rates, poorly educated communities, and communities disconnected from political power.

Bullard [6] documented the patterns of discrimination that concluded minority communities suffered more environmental health problems and experienced higher environmental risks due to the differences in siting hazardous waste facilities within the United States. This pattern of co-locating high risk facilities in communities least likely to have resources to protect themselves not only occurs at national levels but also international levels.

Early research in environmental justice related to the international hazardous waste trade utilized a simple subdivision of distributive justice and procedural justice [11, 39]. Distributive justice deals with the inequalities in the patterns of resources, including political, economic, and social resources [1]. Procedural justice focused on the inequalities in participation in decision-making processes and resource allocation processes [1]. Procedural justice may also be described as fairness.

An inopportune statement by the World Bank Chief Economist Lawrence Summers heightened concerns about international environmental justice. In 1991, Summers stated that the World Bank should encourage the worst polluters to move to the developing South [33]. Summers's memo argued that the South experienced under-pollution due to its lack of economic development, including low wages. Thus, polluting industries should relocate these damaging industries to the South in order to spread out environmental damage and equalize loss of life and other permanently damaging health conditions due to toxic chemical exposure.

Given the lack of resources to adequately manage complex industrial systems, this proposal all but guaranteed that Southern citizens' environmental risk exposure would increase. While the economics of this matter may be sound, the callousness for the suffering of the people exposed to the harmful nature of the toxic waste amplified calls for environmental protection from these practices. The World Bank attempted to provide political cover for this damning memo, but never fully recovered its international prestige.

The United States, via Executive Order 12898 issued by President Bill Clinton in 1994, incorporated evaluating environmental justice concerns into EPA decision-making. This change attempted to eliminate environmental racism by forcing federal agencies to integrate environmental justice considerations into decision-making processes. While the executive order did make federal agencies and the general public more aware of environmental justice concerns, this order did not universally lower the environmental risks or environmental health impacts upon vulnerable communities. As long as the federal agency properly considers the environmental justice considerations, the agency is free to take actions that increase these risks and actual impacts.

This pattern of locating hazardous waste disposals in disadvantaged communities also occurs internationally as well as domestically within the United States [1, 18, 23, 29]. However, scholars disagree about the extent to which environmental racism occurs. Park [23] argued that the presence of the Basel Convention and the Bamako Convention limits claims of environmental racism as African states, in particular, demonstrated the political mobilization to strongly discourage hazardous waste disposal in Africa, even if illegal dumping does sometimes occur. Widawsky [39] agrees and states that the polluter pays principle, and the precautionary principle mitigates some of the characteristics of environmental injustice.

Bullard's pioneering efforts led to a significant expansion of environmental justice research that encompasses the fields of criminal justice, international law, international relations, philosophy, political science, and sociology, along with other areas. Environmental justice research in this time frame represented and continues to represent one of the state-of-the-art approaches to explaining persistent inequalities within international environmental affairs.

6.5   Two-Level Games

The hazardous waste regime may be described as a two-level game [20]. This term refers to a situation in which domestic and international treaties, laws, and regulations interact. The trade of hazardous waste is at once international as the waste crosses one or more boundaries, but also national in that waste management occurs under the domestic laws of a particular country. Thus, transboundary hazardous waste management may be conceptualized using the two-level game concept.

Putnam [25] developed the two-level game theory model of international relations. In this two-level game, domestic interests seek to pressure the national government to adopt their preferred policy outcome. This domestic interest group may vary according to national circumstance but typically represents a group or faction that the national leader relies upon to complete the ratification process or maintain political power.

The national leader engages in bargaining with domestic factions and international counterparts simultaneously, seeking to maximize their domestic power while minimizing harm. Thus, the national leader seeks to identify a win-set. This solution satisfies both the national faction and the international community. International cooperation occurs when the win-sets of both states overlap.

Putnam's two-level game may be widely applicable across multiple international diplomatic events, including hazardous waste trade. Intuitively, Montgomery's [20] statement about the Basel Convention acting as a two-level game appears to fit the evidence. Environmental groups certainly sought to limit the production and movement of hazardous waste. Industrial groups sought to preserve their ability to dispose of waste cheaply. The United States negotiating position, to a certain extent, reflected both of these positions. On the one hand, the United States sought to keep the transboundary movement of hazardous waste open to other countries.

At the same time, however, the United States also saw recovery and recycling as an environmentally sound activity compared to the environmental damage that occurred from extracting virgin raw materials [20].

DeSombre [9] also examined the interactions between the national and international levels. She concluded that domestic regulations might become sources of international environmental policy. She believed that the United States sought to internationalize regulations when doing so presented a clear advantage to its industries and its environmental organizations. The United States seeks to make international standards uniform so that its domestic corporations do not suffer from a competitive disadvantage.

However, the United States also needs an environmental reason for pressing its own environmental regulations to the international level. Within the Basel Convention, the United States succeeded by incorporating RCRA into the treaty. Countries de facto rely on the hazardous waste definitions embedded in this law to define hazardous waste under the Basel Convention.

The usefulness of understanding the linkages between the national and international levels increased over time due to economic globalization. The increases in global communications and transnational movement of goods and services, including hazardous waste necessitated more control over this movement. The Basel Convention, its protocols and amendments, and other important regional treaties that regulate the hazardous waste trade provide significant protection from improper disposal of these toxic wastes.

References

1. Anand R (2004) International environmental justice: a North-South dimension. Routledge, New York
2. Basel Convention, 22 Mar 1989. https://treaties.un.org/doc/Treaties/1992/05/19920505% 2012-51%20PM/Ch_XXVII_03p.pdf
3. Bradford M (1997) The United States, China & the Basel convention on the transboundary movements of hazardous wastes and their disposal. Fordham Environ Law J 8(2):305–349
4. Bruno K (1998) Philly waste go home. Multinatl Monit 19(1–2):7–9
5. Bullard RD (1983) Solid waste sites and the black Houston community. Sociol Inq 53(2–3):273–288
6. Bullard RD (1993) The threat of environmental racism. Nat Resour Environ 7(3):23–56
7. Choksi S (2001) The Basel convention on the control of transboundary movements of hazardous wastes and their disposal: 1999 protocol on liability and compensation. Ecol LQ 28(2):509–540
8. Clapp J (1994) The toxic waste trade with less-industrialised countries: economic linkages and political alliances. Third World Q 15(3):505–518
9. DeSombre ER (2000) Domestic sources of international environmental policy: industry, environmentalists, and US power. MIT Press, Cambridge
10. Gibbs LM (2011) Love Canal: and the birth of the environmental health movement. Island Press, Washington, D.C.
11. Gutierrez R (2014) International environmental justice on hold: revisiting the Basel ban from a Philippine perspective. Duke Environ Law Policy Forum 24(2):399–426
12. Heiman M (1990) From "not in my backyard!" To 'not in anybody's backyard!' J Am Plann Assoc 56(3):359–362

13. Horinko ML (2002) 25 years of RCRA: building our past to protect our future. Office of solid waste and emergency response EPA-K-02-027, EPA, Washington D.C.
14. Jaffe D (1995) The international effort to control the transboundary movement of hazardous waste: the Basel and Bamako conventions. ILSA J Intl Comp L 2(1):123–138
15. Kellenberg D, Levinson A (2014) Waste of effort? International environmental agreements. J Assoc Environ Resour Econ 1(1/2):135–169
16. Krueger J (1998) Prior informed consent and the Basel convention: the hazards of what isn't known. J Environ Dev 7(2):115–137
17. Kummer K (1992) The international regulation of transboundary traffic in hazardous wastes: the 1989 Basel convention. Int Comp Law Q 41(3):530–562
18. Marbury HJ (1995) Hazardous waste exportation: the global manifestation of environmental racism. Vand J Transnatl L 28(2):251–294
19. McKee DL (1996) Some reflections on the international waste trade and emerging nations. Int J Soc Econ 23(4/5/6):235–244
20. Montgomery MA (1990) Traveling toxic trash: an analysis of the 1989 Basel convention. Fletcher Forum World Aff 14(2):313–326
21. Okaru VO (1993) The Basel convention: controlling the movement of hazardous wastes to developing countries. Fordham Environ Law Rev 4(2):137–165
22. Paigen B (1982) Controversy at Love Canal. Hastings Cent Rep 12(3):29–37
23. Park RS (1998) An examination of international environmental racism through the lens of transboundary movement of hazardous wastes. Indiana J Glob Leg Stud 5(2):659–709
24. Pellow DN (2007) Resisting global toxics: transnational movements for environmental justice. MIT Press, Cambridge
25. Putnam RD (1988) Diplomacy and domestic politics: the logic of two-level games. Int Organ 42(3):427–460
26. Rublack S (1989) Fighting transboundary waste streams: will the Basel convention help? Verfassung Und Recht in Übersee 22(4):364–391
27. Schmidt CW (1999) Trading trash: why the U.S. won't sign on to the Basel convention. Environ Health Perspect 107(8):A410–A412
28. Schwartz J (2000) Garbage ash that nobody wanted. The Seattle Times. https://archive.seattl etimes.com/archive/?date=20000917&slug=4042847. Accessed 16 Dec 2021
29. Simon DR (2000) Corporate environmental crimes and social inequality: new directions for environmental justice research. Am Behav Sci 43(4):633–645
30. Shearer CRH (1993) Comparative analysis of the Basel and Bamako conventions on hazardous waste. Envtl L 23(1):141–183
31. Strohm LA (1993) The environmental politics of the international waste trade. J Environ Dev 2(2):129–153
32. The Economist (1989) The garbage industry: where there's muck there's high technology, p 23
33. The Economist (1992) Let them eat pollution, 8 Feb, p 82 (U.K. edition)
34. UNEP (1982) Montevideo programme for the development and periodic review of environmental law. UNEP/GC/DEC/10/21
35. UNEP (1987) Ad Hoc working group of legal and technical experts with a mandate to prepare a global convention on the control of the transboundary movements of hazardous wastes report of the meeting, organizational meeting, Budapest 27–29 Oct 1987 report of the meeting. UNEP/WG.180/3
36. UNEP (1988) Ad Hoc working group of legal and technical experts with a mandate to prepare a global convention on the control of the transboundary movements of hazardous wastes report of the meeting report of the working group Geneva, 1–5 Feb 1988. UNEP/WG/.182/3
37. UNEP (1994) Second meeting of the conference of the parties to the Basel convention on the control of transboundary movements of hazardous wastes and their disposal. UNEP/CHW.2/30
38. United Church of Christ Commission for Racial Justice (1987) Toxic wastes and race in the United States: a national report on the racial and socio-economic characteristics of communities with hazardous waste sites. https://www.nrc.gov/docs/ML1310/ML13109A339.pdf. Accessed 16 Dec 2021

39. Widawsky L (2008) In my backyard: how enabling hazardous waste trade to developing nations can improve the Basel convention's ability to achieve environmental justice. Envtl L 38(2):577–626
40. Yakowitz H (1987) Waste management activities in selected industrialized countries. OECD Environment Directorate, Paris

Chapter 7
The Earth Summit and Its Aftermath

Abstract State support for sustainable development emerged in the aftermath of the World Commission on Environment and Development's 1987 report *Our Common Future*. Coinciding with the end of the Cold War, states shifted their focus to economic and social affairs, including environmental problems. The decision to celebrate the twentieth anniversary of the United Nations Conference on Human Environment launched preparations for the United Nations Conference on Sustainable Development, the second of the environmental mega-conferences. States quickly negotiated treaties dealing with biodiversity and climate change and produced Agenda 21, a blueprint to implement sustainable development. Additionally, visibility of the role of NGOs within international environmental diplomacy changed during this mega-conference. States did not complete their ambitious agenda for this meeting as a treaty about forestry failed to materialize. However, scholars and participants alike believe the conference succeeded in promoting sustainable development.

Keywords Earth Summit · United Nations Conference on Environment and Development · Sustainable development · Non-governmental organizations · International institutions

In the twenty years since the UNCHE meeting in 1972, environmental affairs occupied a curious place on the international agenda. While the issue originally arrived with great fanfare in Stockholm, the energy and sophistication brought to the topic proved difficult to sustain. The UN system that had built momentum for increased environmental protection through its quick action in creating UNEP watched the system become mired in technical controversy as quick victories gave way to disappointment as negotiations stalled for a variety of reasons, including a lack of funding and scientific uncertainty [17, 25]. The tenth anniversary of the UNCHE conference, marked by The Nairobi Session of a Special Character in May 1982, should have been a time of celebration. Instead, diplomats bemoaned the decline in enthusiasm [25]. The seeming lull in international environmental negotiations hid a great deal of environmental management and consensus-building. Compared to the frenzied activity that occurred as the 1980s drew to a close, the activity in the early part of the decade paled in comparison.

At the same time, however, the end of the Cold War in the late 1980s did significantly impact global politics. By 1992, political thought believed that the end of the Cold War left open the possibility of finally creating a "new world order" based upon cooperation between nations. In the absence of hard power politics such as military and economic standoffs, states could focus their attention on the softer side of international order, including environmental affairs. German reunification seemed to suggest that the European divide between East and West would cease to exist and that the UN could turn its attention to other areas of importance. Southern states hoped that the end of the Cold War would create an opportunity for their economic development.

While world events related to the end of the Soviet Union continued to dominate the international news, including the breakup of Yugoslavia, the most visible change marking the end of the Cold War involved changes in physical territory. Areas under the control of the former Soviet Union wasted little time expressing their desire to join the EU or at least claim neutrality. The waning influence of the Soviet Union brought about the end of the Soviet occupation and brought hope that the problems that consistently plagued humanity, hunger, poverty, and economic injustice, might move into the limelight. Southern countries hoped that resources once devoted to military and economic domination could be reassigned to address these lingering problems that impacted people both inside and outside the confines of the American and Soviet locales.

Thus, the end of the East–West conflict created a void in the international agenda that allowed the North–South gap to move to the forefront of international politics [2]. Accordingly, the developing countries successfully used their dominant voting power within the UN GA to focus global attention on their needs for socioeconomic development [12]. The rise of sustainable development that coincided with the end of the Cold War provided a convenient ideological background to promote this most recent articulation of eco-development.

This chapter reviews the UNCED conference beginning with the articulation of sustainable development in 1980 through the arrival of diplomats in Rio de Janeiro in the summer of 1992. Section one begins with the articulation of sustainable development in the 1980 World Conservation Strategy. Section two provides an overview of the Earth Summit where diplomats succeeded in creating a watershed moment for international diplomacy [18]. Section three concludes the chapter by examining the legacy and aftermath of this conference.

7.1 Organizing the Conference

Historical investigations into the creation of the sustainable development paradigm typically point to the 1980 World Conservation Strategy authored by the International Union for Conservation of Nature at the bequest of UNEP [13]. In all probability, the intellectual origins of sustainable development began before its introduction in the 1980 World Conservation Strategy [8]. At least one NGO, the Environment Centre

Liaison International, claimed to have worked on the conceptualization of sustainable development in 1976 and 1977 [10].

By the late 1970s, progress on creating new international environmental treaties stalled with little to no progress on key issues such as ozone depletion or protecting the marine environment. While the UNEP Governing Council sought ways to create a more robust environmental agenda through the Montevideo Environmental Law Program, an easy victory for environmentalists did not appear in the works. After a lackluster review of the implementation of the Stockholm Action Plan in 1982, the UN GA formed the World Commission on Environment and Development (WCED), charging the group with investigating the human environment and interlinkages to economic development. This group, more commonly known as the Brundtland Commission, after its chairwoman Gro Harlem Brundtland, published its final report in 1987.

WCED convened a three-year work plan to host public workshops to gather information about the relationship between environment and development. The twenty-three-member commission represented diverse views about the relationship between environment and development. Borowy [5] remarked that WCED did more than provide an academic definition of sustainable development, the context in which the report is most often cited. This commission also intended to create international environmental policy change by soliciting new ideas about sustainable development and making policy recommendations for all international actors.

One of the many international policy actions recommended by the Brundtland Report was a global conference on environment and development. The publication of *Our Common Future* called for a worldwide agenda based upon the need to focus on people's economic, environmental, and social equality. Consequently, states responded by initiating a second global gathering by strengthening the ties between environment and development, a goal never attained by the proponents of eco-development discussed in Chap. 2. This report also provided the most often cited definition of sustainable development as "development that meets the needs of the present without compromising the ability of future generations to meet their own needs" [33: 54].

Acting upon this advice, the UN GA passed Resolution 44/228 on December 22, 1989. The resolution began preparations for UNCED, more commonly known as the Earth Summit, and accepted Brazil's offer to host the meeting in 1992. Thus, environmental affairs ascended to the top of the global agenda, a mere twenty years after it first arrived on the international stage. While this time period may seem long to global citizens, diplomatic wisdom suggests that international environmental issues had barely reached their maturity.

As with any major event, one of the first decisions revolves around which organization will take the lead in organizing the conference itself. In this case, member states decided that the UN GA would directly oversee the conference organization. This decision ensured the conference agenda and logistics would be closer to the interests of the numerically superior Southern block of countries. Thus, Southern interests assumed primacy, meaning that development concerns could potentially triumph over limiting or reversing environmental damage caused by industrialization. However, this decision also dealt a blow to the prestige of UNEP as the shift to the UN GA highlighted states' perception of UNEP as a small, highly technical agency with limited political power to oversee a conference of this size and potential magnitude. This view is not surprising given that these same states designed UNEP with limited capability to play a strong political role within the UN.

In an interesting twist of fate, Canadian Maurice Strong, the energetic Secretary-General at UNCHE twenty years earlier (and the first head of UNEP), was appointed by the Secretary-General of the UN, Pérez de Cuéllar, to serve as Secretary-General of the Earth Summit in February 1990. In doing so, de Cuéllar increased the political legitimacy and authority of the Rio conference by increasing global expectations for a conference that would improve the condition of the global environment. In the twenty years since the UNCHE conference, Maurice Strong had solidified his international reputation as a unique voice within international environmental politics. On the one hand, Strong was a Canadian businessman with deep ties to the oil and gas industry, responsible either directly or indirectly for many of the ills on the agenda at both conferences. On the other hand, Strong's reputation as an international humanitarian with a great concern for the impoverished countries and people around the world won him many accolades and international political support. Chairing a second major conference on global environmental issues also strengthened Strong's legacy as one of the leaders of the world environmental movement.

The preliminary process for UNCED consisted of four meetings held in five locations. The first, in New York, organized the process, and the other four dealt with more substantive issues. The initial Organizational Meeting was held from March 5–16, 1990, and Tommy Koh of Singapore was elected as Chair. The structure of the PrepCom was determined to include three working groups and a plenary session. The Conference Secretariat staff located in Geneva provided logistical support to this group. These individuals also provided the first draft of many of the documents discussed during the meetings. Conference Secretariat staff could discreetly shape the course of work through this responsibility, although reports had to conform to national governments' express guidelines and implicit expectations [34]. Conference rules dictated the adoption of documents by consensus. This procedural mechanism effectively gave each state a veto over any specific word (or group of words) in the documents. Established diplomatic tradition utilized "bracketed text" to signal dissent. Conference rules did not provide provisions for voting.

This meeting also created committees to begin the work of negotiating outcomes. Working Group I focused on atmospheric issues (i.e., climate change, stratospheric ozone, and transboundary air pollution), biodiversity, and biotechnology [30]. In contrast, Working Group II focused on the protection of water (i.e., oceans, seas, coastal areas, and freshwater resources), waste, toxic chemical management, and transportation of hazardous materials. Additionally, the role of NGOs became a controversial item during this time frame. Southern countries, deeply suspicious about environmental groups derailing Southern development concerns, initially sought to limit NGO access to the PrepCom processes [35]. This limitation stood in stark contrast to past UN conference traditions that allowed NGOs registered with the

ECOSOC to actively participate in the preparatory phases of conferences. Indeed, some precedence existed, including from UNCHE in 1972, that NGOs that were not registered with ECOSOC could also participate in the conference, assuming no significant objection from the chair of the meeting occurred. The resistance to NGO access was so significant that the action item was forwarded to the next PrepCom in Nairobi.

The second half of PrepCom I occurred August 6–31, 1990, in Nairobi, with the main focus to determine the conference topics. PrepCom I determined that the environmental agenda would not overshadow developing countries' developmental priorities. Member states conducted negotiations for the Earth Summit beginning with a review of international environmental diplomacy [29]. As conversations for UNCED began, diplomats formed an agenda to consider environmental problems such as climate change, biodiversity, forestry, land-based pollution, hazardous waste problems, and water-based pollution, including oceans and regional seas [30]. In the past, environmental concerns of the North tended to trump the developmental concerns of the South, a fact of international affairs that continues today. The working agenda then split into six organizational areas (1) Conventions; (2) Earth Charter; (3) Agenda 21 (an action plan to achieve sustainability); (4) Financial resources; (5) Technology transfer; and (6) Institutions [28].

The effects of the Southern negotiating bloc impacted the UNCED process as climate change, biodiversity, forests, and land-based pollution, all traditional Northern concerns, were grouped into one single agenda item. Financial resources and technology transfer, two distinctly Southern concerns, became "new" items. These two concerns indicate the two primary thrusts of development aid and assistance from North to South. Financial resources would assist Southern countries with improving their financial health as it involved the conditions under which these states accessed financial resources, including monies via loans, donations, and grants.

Technology transfer represented future economic growth by allowing Southern countries access to technology that would allow their industrial hubs to compete more evenly with the industrialized North. The disagreement between North and South did not stem from the need for technology transfer but rather the terms and conditions for the transfer. The South insisted on technology transfer at a steeply discounted rate, or preferably for free. The North disagreed and thought that technologies should be transferred at market cost. In other words, the South wanted the North to give away their patented technology far below market costs, and the North offered their intellectual property for sale as a regular commercial transaction.

PrepCom II was held in Geneva from March 18–April 5, 1991, and this session focused primarily on financial resources and technology transfer [11]. Unfortunately, the Earth Summit occurred during a global recession, and access to new developmental assistance did not meet developing countries' needs or desires. Equally important, the United States argued against technology transfer for the simple reason that private companies owned the technology. The United States does not have the legal ability to order private companies to give their business trade secrets away.

As in most international negotiations, the second meeting did not yield substantive progress on draft decisions. However, much discussion occurred that would, in due course, provide the basis for conferees to agree to a substantial amount of text in advance of the conference. Adams and Martinez-Aragon [1] report that delegates failed to make much substantive progress during PrepCom II, except for the text on forestry. As a result, pressure mounted on delegates to move forward with finalizing texts during the third and fourth meetings.

PrepCom III was also held in Geneva from August 12–September 4, 1991. The three working groups did not make any substantial progress as Working Group I prioritized discussing energy policy while Working Group II prioritized oceans. During this meeting, diplomats realized that a convention on forests would not be forthcoming, and ambitions to create this new international treaty downgraded into a non-binding statement on forestry conservation and management.

More than any substantial progress, this session was noted for its innovations in meeting participation. PrepCom III saw the introduction and differentiation of the "formal–informal" and the "informal-informal" meetings. "Formal–informal" meetings typically allowed NGO access and language translation, while "informal-informal" were held in a private area and excluded NGOs. Additionally, the sessions were conducted exclusively in English, a stark departure from diplomatic tradition that typically provided translators, potentially disadvantaging small developing countries that did not speak English. Notably, NGO access during PrepCom III depended upon both the countries involved and the topic being discussed as the Working Group Chair had the authority to admit or exclude NGOs. While the majority of states recognized the need to provide more transparency and consequently more access to the meetings, states nevertheless felt that some issues were too sensitive for NGOs, such as Working Group III that dealt with the legal status of treaties.

While not part of the formal negotiating processes, but nevertheless profoundly impacting the environmental arena, the World Bank, beginning in 1989, organized action to create an environmental fund based upon a guarantee of funds from France. Officially overseen by the World Bank, UNEP, and UNDP, the Global Environment Facility (GEF) began as a three-year pilot program in October 1991. Other developed countries also contributed funds that could underwrite projects in developing countries that implemented environmentally beneficial projects. GEF focused on four areas, including the recently signed Montreal Protocol, protection of international waters, and the nascent program areas of climate change and biodiversity.

PrepCom IV was easily the most intense of the preparatory meetings. It varied from the traditional format in that five weeks, from March 3–April 3, were dedicated to discussions in New York. The meetings frequently ran around the clock, and the final session closed in the pre-dawn hours on April 4. PrepCom IV saw the completion of many of the documents forwarded to the Earth Summit and near completion of many other important documents, including the Rio Declaration and Agenda 21. Fletcher [9] reported that delegates had finished negotiations on 90% of the texts for the Summit, but documents regarding financial resources, technology transfer, and forest principles remained incomplete. This level of completion of the final texts increased the probability for a successful conference as this allowed the heads of state to identify areas where political compromise was needed instead of the conference text suffering from lack of resolution of technical issues.

Bernstein et al. [4] disagree. They believe that states put off many of the more controversial items at PrepCom IV, with the net impact of transforming the main conference from a thinly-veiled photo opportunity to a substantive negotiating session. This observation is noteworthy as it came via the Earth Negotiations Bulletin (ENB), a newly created newspaper of sorts that recapped major conversations during the day along with a brief commentary on the significance of the negotiations themselves. Indeed, the ENB itself made a significant contribution to international environmental diplomacy. Today, the newsletter is published for every negotiating session of importance, with past issues archived online.

While international environmental diplomacy traditionally emphasizes the role of states, other groups also have a vested interest in the outcomes of the negotiations. The UN has been more willing to allow access to non-state observers within international environmental diplomacy than other issue areas. Along with detailing the positions of various states during the negotiations, the ENB reports document the many places where NGO input directly impacted the negotiations. Equally important, it is likely that the impact of NGOs was much broader reaching than the issues mentioned in the ENB, as evidenced by the introduction of "formal-formals" and "informal-informals" meetings at the PrepCom. There would be no need for this innovation if governments were not reacting to the presence and input of the NGOs.

In all likelihood, Strong's experiences from Stockholm some twenty years earlier influenced his decision to involve NGOs in the process to lower the possibility of ugly confrontations and protests. Strong's vision of NGOs included their knowledge of environmental and developmental problems and their capability to publicize and legitimize the entire Rio process. For this vision to be successful, NGOs needed access to the negotiations or at least the promise of access. A UN GA Resolution (A/RES/46/168) asked the Conference Secretariat to invite all NGOs that had been accredited to the PrepComs, in addition to the 178 national delegations. However, the Rules of Procedure differed from the PrepComs as NGOs could address the Conference only at the invitation of the presiding officer and with delegate approval. They were, however, able to distribute written statements without having to secure formal permission.

7.2   The Earth Summit

Diplomats met from June 3–14, 1992. The Conference was organized into two sections. The first constituted a lower-level technical negotiating session that lasted for ten days. The second section consisted of the high-level "Summit" and lasted for two days. For the most part, heads of governments and official delegations approved the prenegotiated documents. Texts that states forwarded to Rio de Janeiro with bracketed texts were generally finalized before the Summit. This is typical for international conversations as diplomats very rarely complete all the negotiations before the actual conference as highly political topics may require input directly from a head of state. Thus, while much of the real negotiations occur beforehand, the remaining text frequently generates significant concern and controversy.

Major agenda items included the two treaties, the UN Framework Convention on Climate Change (FCCC) and the CBD. The first treaty dealt with nascent attempts to quantify and limit the growth of greenhouse gases in the atmosphere, while the second treaty sought to preserve our natural resources and promote sharing the benefits of goods produced from genetic resources. Both of these treaties are covered individually, with climate change in Chap. 8 and biodiversity in Chap. 9. Additionally, diplomats produced a statement of forest principles, Agenda 21, and the Rio Declaration on Environment and Development. Unlike the two predecessor treaties, these topics generated widespread resistance and controversy, leading to dismal prospects for future implementation.

The Statement of Forests arose out of a convention initiative proposed by President George H. W. Bush at the July 1990 Economic Summit of the G-7. The original work plan for the conference envisioned a binding treaty on forests. However, developing countries, led by Malaysia, combined with powerful US business interests and successfully resisted international intrusion into their internal affairs [1]. They insisted the treaty unfairly singled out certain types of forests and was thus an encroachment of national sovereignty. Instead, a last minute compromise brought forth the Statement of Forest Principles. This non-binding statement provided guidelines for the conservation and sustainable management of forests by equalizing economic development and preservation of forests for traditional uses by indigenous peoples and maintaining biological diversity.

Originally designed to be "The Earth Charter," delegates renamed the document the Rio Declaration on Environment and Development after disagreeing on an acceptable philosophical approach. The Earth Charter was the brainchild of Maurice Strong, who hoped to establish the ideological guidelines for protecting the environment. However, states failed to reach a consensus on these underlying guidelines, as the North–South gap proved too wide to reconcile. As a result, diplomats drafted the Rio Declaration in a face saving measure. The Rio Declaration contained 27 principles on sustainable development intended to commit states to work toward the laudable goal of equitable partnership and cooperation to protect the earth.

Secretary-General Strong intended Agenda 21 to be the policy blueprint to guide the transition to a more equitable, environmentally friendly global socioeconomic system. The 40 chapters in Agenda 21 roughly fall into four sections: social and economic dimensions, environmental resource management, the role of major groups (including NGOs), and implementation [26]. However, Chap. 4, dealing with consumption, was not widely accepted, with the United States leading the opposition to the document as President Bush reportedly declared that "The American way of life is not up for negotiation" [26].

Agenda 21 devotes Chap. 27 to NGOs. In it, the UN declares that NGOs are an essential part of participatory democracy [28]. Additionally, the UN commits itself to involve NGOs within the UN system, including mandating institutions to work more closely with this sector. It also suggests that NGOs will need to "foster cooperation and communication among themselves to reinforce their effectiveness as actors in implementing sustainable development" [28: 27.4]. Agenda 21 also called for a review no later than 1995 of how NGOs participate within the UN system, namely the ECOSOC registry.

Member states also discussed desertification, particularly in Africa, as part of the Rio Conference. Deliberations about desertification centered on whether or not an international negotiating committee should convene. The Earth Summit recommended that the General Assembly established this committee with a June 1994, deadline for producing consensus on a treaty text. The UN GA gave its approval in late December 1992, and the INC to Combat Desertification met five times before successfully concluding treaty negotiations [31]. Further, items such as military and the environment, linkages between free trade and the environment, biotechnology, and international enforcement activities were conspicuously absent from the agenda to the disappointment of the more radical wing of the environmental movement.

Diplomats did agree, however, to support the newly created GEF [11]. While Southern countries, along with NGOs, viewed the GEF with significant suspicion, Northern countries adamantly opposed creating new funding sources for environmental purposes [23]. The GEF, in its current configuration, was controlled by Northern countries through the World Bank model that prioritized donor countries. Southern countries preferred utilizing a governance model based upon the UN system, where the Southern country would be guaranteed political control. Consequently, utilizing the GEF for the newly created climate change and biodiversity regimes would be contingent upon GEF reform after the Earth Summit.

Intergovernmental organizations were not the only groups seeking to participate alongside states at UNCED. NGOs had their largest encounter to date with the UN system at UNCED. The Conference established new practices regarding access to the UN negotiating table. It recognized the capabilities of NGOs to promote environmental values and political support for UN initiatives more generally. These contributions culminated in a parallel conference that attracted roughly 20,000 participants [21]. At least three NGOs were created specifically to help with organizing non-state actor participation at the Earth Summit, the International Facilitating Committee (IFC), ECOFUND'92, and the Business Council for Sustainable Development (BCSD) [32].

The conference staff gave preferential treatment to the IFC established under the auspices of the Center for Our Common Future [8]. However, disagreements within this group broke out in 1991, and the NGO community splintered into multiple organizing groups, nominally along goals and ideology. Environmental NGOs interacted cooperatively with each other, while business and industry groups tended to collaborate with other business and industry organizations. In addition to differences in agendas, NGOs argued among themselves about who should be allowed to attend UN meetings as NGOs. NGO objections were particularly strong toward the BCSD and the transnational corporations they represented.

Not surprisingly, prominent groups organized separately from the IFC-led process, including business NGOs and environmental groups politically opposed to the IFC leadership. Developmental and environmental NGOs partially merged their agendas in keeping with sustainable development themes. Third World Network, headed by

Martin Khor, was particularly effective in highlighting the South's perspective on environment and development. One method utilized by this NGO included releasing a series of briefing papers highlighting the differences between North and South. As a result of this effort and others like it, northern NGOs would no longer be able to talk about environmental issues solely in terms of wildlife issues or population. Instead, they would incorporate into their debate issues such as social justice and environmental equity.

The Brazilian Forum of NGOs, on behalf of NGOs and other interested civil society groups, organized the Global Forum. The Global Forum was a series of planned and spontaneous events to raise public awareness of the Earth Summit. It was designed, in part, to relieve some of the NGO pressure at the official conference. Bernstein [3], writing to brief the Canadian delegation, expected that NGO demand for observer status would exceed the space available. As a result, NGOs that were not accredited to PrepCom IV would not be allowed to attend the formal conference. Instead, NGOs were expected to attend the parallel forum. Bernstein expected that over 200 events involving 3000 NGO participants would occur as part of the Global Forum. Instead, the Global Forum produced over 300 events, and attendance at the Flamingo Park exceeded half a million people for the two-week duration of the conference.

The Global Forum produced forty-six "treaties" and statements on diverse topics such as education, communication, cooperation, and a code of conduct for NGOs, racism, militarization, natural resource use, climate change, biodiversity, and biotechnology. The NGO Earth Forum database provided further information by listing the signatures for thirty-six of these treaties. Additionally, meetings, workshops, and mock parliamentary sessions were held, movies were shown, and position papers were produced and discussed [35]. In short, the Global Forum became one of the largest networking and information gathering meetings on environment and development to date.

An essential skill at the UN conferences is assembling and then maintaining coalitions among like-minded states [34]. NGOs are not an exception to this expectation. Quite the opposite, UN diplomats and state negotiators expect NGOs to assimilate their views into broad-based coalitions. NGOs attending Rio needed to organize and prepare a platform to do so. These two characteristics combined give business and industry NGOs a tactical edge with respect to the UNCED process. However, large environmental NGOs such as IUCN, WWF, and Friends of the Earth also garnered significant attention. The remainder of the NGOs did not seem to have a coherent message. Their differences in opinion with respect to analysis and preferred solutions could not be presented as a joint statement, a distinct disadvantage when dealing with the UN.

Despite this impressive list of achievements for a single conference, attendees at the Rio Summit left with a mixed feeling about the meeting. Strong [24] recalls in his autobiography that he gave considerable thought to declaring the meeting a failure as it fell short of his expectations. Fletcher [9] reported G-77 disappointment when governments failed to financially support Agenda 21 in the 1992 fall UN GA meeting. In the end, however, rhetoric at the meeting tended to acknowledge the successes of the Earth Summit while glossing over its shortcomings.

7.3 The Earth Summit Legacy

What then is the legacy of the Earth Summit? It is undoubtedly safe to say that many if not all, countries were interested in the twin issue areas of environment and development at the time of the conference itself. Indeed, one feature of contemporary society is that global interest in both environment and development, while moving through cycles of interest and disinterest, never left the international agenda. To take a closer look at the legacy of the Earth Summit, the remainder of the chapter looks at key themes that emerged because of this meeting. These themes include evaluating the conference's success, the meshing of environment and development into sustainable development, the role of non-governmental organizations, and the complexity of the international environmental agenda.

Today, the Earth Summit is heralded as the pinnacle of success within international environmental diplomacy. While it is undoubtedly true that states failed to support either Agenda 21 or the Rio Declaration vigorously, certainly the UNFCCC and the CBD became major international environmental frameworks in the twenty-first century. One key line of reasoning to consider while analyzing the legacy of the Earth Summit is to decide the meaning of success for international environmental diplomacy. Differing definitions of success exist in line with the multifaceted nature of the multilateral conferences themselves. Scholarly analysis of conferences details the conferences by the numbers. This methodology tempts untold numbers of students and scholars due to the simplicity of counting the number of states that sign a treaty, the number of participants that attend a meeting, or the number of treaties that occur. Ranking the Earth Summit by any of these statistics places the Earth Summit as one of the most successful conferences held to date. UNCED holds the record for the largest number of participants and the largest number of treaties to emerge at a single conference. Recalling from Chap. 5, the Montreal Protocol to Protect the Ozone Layer might be seen as the premier treaty as it is the only one universally agreed to by all countries of the world.

International relations scholarship, including regime theory, focuses on success in terms of international structures [36]. In other words, international environmental treaties may be measured as successful treaties when the institutions they create function as designed. In this instance, the UN GA, implementing a recommendation from Agenda 21, created the Commission on Sustainable Development (CSD) to ensure follow through with the commitments made during the Earth Summit. The CSD was established with the laudable goal of making political decisions about environmental policy and overseeing the coordination of environmental affairs within the UN system along with UNEP. Instead, the CSD settled, somewhat uncomfortably, into its role as a yearly debating forum for major stakeholders.

Dimitrov [7] instead focuses on international diplomatic failures utilizing the CSD as an example. This empty institution occurred because diplomats have incentives to create the illusion of cooperation and activity, even when states do not want the activity to produce results. In other words, institutions can be designed in such a way as to deliberately foreclose meaningful results.

Alternatively, Kütting [14, 15] argues that an eco-holistic approach that links environmental treaties to improved environmental conditions should be the measure of success. However, relating environmental treaties to actual environmental improvement proves to be difficult due to the inability to link cause and effect. While it may well be possible to point to how a treaty on hazardous waste has reduced the toxicity of shipments into a specific country, other cases may be much more difficult to assess. For example, how does one assess Agenda 21's Sect. 21.4 that attempts to change consumption and waste patterns using life cycle analysis to changes in the distribution of landfills that hold the unwanted hubris of a society [26]?

Another aspect of determining the success (or failure) of the Earth Summit should evaluate the impact of the sustainable development norms on human history. As shown in previous chapters, one of the primary theories of international environmental diplomacy states that changes in norms cause changes in the behavior of key actors. Thus, one way to judge the success of the Earth Summit could be to look at the new norm of sustainable development and how this norm has spread. Changes in behaviors could be linked to the norms by surveys or other measures that evaluated sustainable development project implementation. However, given that twenty-plus years have passed since the Earth Summit, questions of causality would arise in showing that the actions taken in support of sustainable development were due to the Earth Summit (and only to the Earth Summit).

Equally worthy of consideration is the point that decisions made at international conferences influence the future work agendas of all kinds of international actors, including states, businesses, NGOs, civil society, and individuals located around the globe. In other words, influence and impact do not constitute a one-way street. While one of the primary points of this book is that many individuals working together impact the outcome of international environmental diplomacy, the formal and informal decisions reflected in the negotiated text also impact individuals and the groups that they form within our global society.

While it would be accurate to remember the Earth Summit as a qualified success, the Earth Summit also altered the future trajectory of international environmental negotiations. In 1992, Rio installed sustainable development as an integral, although separate, part of the international environmental agenda. Srinivas [22] summarized the purpose of UNCED as the first attempt to make international public policy on the environment and development. This early work defining sustainable development does not entirely match the meaning of the term as it would be used in contemporary contexts in the twenty-first century. Ironically, the diplomatic maneuvering during the Earth Summit initially moved the concept of sustainable development closer to the Northern viewpoint of science-oriented environmental protection over the Southern concerns about development for impoverished countries.

Consequently, Northern sensibilities about lifestyle issues, such as consumption, prevented the developing countries from advancing an agenda that hampered the developed countries from continuing to engage in a highly consumptive lifestyle. There would not be a reduction in consumption to fund Southern ambitions to industrialize at the expense of the North. This consumption issue also meant that waste disposal streams from manufacturing processes or disposal of throw-away items would remain constant in the future.

From a Southern perspective, the Earth Summit fell short of the needs and hopes of the Southern countries [22]. Perhaps understandably, Southern critics viewed the entire dynamic with cynicism and skepticism. During the aftermath of the Earth Summit, the G-77 bloc proposed new programs for developmental aid based on the Earth Summit commitments. Still, they could not force the developed countries to follow through with cash or other valuables. Further, the G-77 bloc extended significant energy, ensuring that much-needed resources would be focused on the developing countries rather than on the so-called Economies in Transition in Eastern Europe that resulted from the breakup of the Soviet Union.

Thus, the Earth Summit marks a pivotal moment within international environmental diplomacy. Treaties negotiated before this point tend to be technical in nature. Certainly, the Montreal Protocol and the Basel Convention focused on specific environmental harms with known causes. The great beauty of these treaties is the ability of humankind to propose solutions, typically in the form of halting environmentally destructive behavior such as limiting the damage from ozone depleting substances or ending the toxic trade. The treaties finalized at and after the Earth Summit may also contain technical issues, but they also include normative elements that add to the complexity of the international agenda. Neither climate change nor biodiversity treaties could be considered an exclusively technical treaty.

One of the lasting structural shifts of the Earth Summit conference included NGO participation and deepening the integration of international environmental affairs within the realm of international governance [6]. These objectives had slowly emerged over the intervening two decades. An increasing world environmental consciousness led to a slow expansion of the international agenda in the two decades between Stockholm and Rio. These issues did not appear suddenly on the global agenda but instead developed during many international meetings, informal conversations, symposia, and workshops. In contrast, Willets [35] points out that many of the so-called NGO innovations highlighted during the Earth Summit occurred ten years earlier at the UNCHE covered in Chap. 2. On the surface, this is true. However, the innovations pioneered by NGOs did not become durable, permanent features of the international structure until the Earth Summit occurred [22].

While NGOs had substantial differences of opinion that resulted in vastly divergent policy suggestions and techniques at Rio, the majority of NGOs supported an increased role for NGOs in implementing sustainable development. The bottom-up approach in Agenda 21 dictates wider participation at the grassroots level. All NGOs supported their individual inclusion in the UN system, regardless of what they thought about their NGO counterparts. Thus, the call for reform in Agenda 21 resulted in a significant expansion of the NGO roster.

However, NGOs did not form into one negotiating block on this issue. The combined efforts of all the NGO groups convinced negotiators to rely upon their mobilization at the local level. In the future, NGO success would depend, in part, on their capability to deliver upon this promise of widespread public participation and support for UN initiatives. The need to aggregate viewpoints to successfully address the main assemblies at Rio contributed to the further politicization of NGOs that started in Stockholm. NGOs now had the opportunity to build coalitions for and against specific policies and inject these viewpoints directly into the UN negotiations either by means of written statements, side-bar events, significant group discussions, or direct statements at the main plenary.

As with any major conference, numerous countries implemented sustainable development as specified in the Rio documents. National workshops and local Agenda 21 Committees were launched throughout the world. That is not to say that states universally adopted sustainable development. Based on a cross-national comparative project of the uptake of sustainable development in nine developed countries and the European Union, Lafferty and Meadowcroft [16] report that states vary significantly in their support of the concept and range from the "enthusiastic" support of Netherlands, Norway, and Sweden to the "disinterested" United States.

However, activity on sustainable development is not limited to governmental authorities. NGOs and businesses have also been active on both sides of this issue. For instance, corporations formed their own NGOs, such as the BCSD, to promote business activity supporting sustainable development. Business and Industry Non-governmental organizations (BINGOs) implicitly understood that decisions taken at the Rio conference would eventually affect individual corporations via new policies at the national level. The importance of the formation of the BCSD lay in its acceptance of the sustainable development regime and inherently intertwined with that, sustainable development norms. Stephen Schmidtheiney, the first president of the BCSD, published a book entitled *Changing Course* that coined the term "eco-efficiency" [20]. Robins [19] states that BCSD consulted closely with IIED in developing the content for BCSD activities and that this initial work led to Lloyd Timberlake (of IIED), who earlier worked on the Brundtland report, serving as a "ghost-writer" for the final draft of *Changing Course*. Eco-efficiency stood for the efficiency improvements that come from doing more with less, i.e., increasing production efficiency through process design changes, including reducing raw materials and energy.

BINGOs have a long and storied tradition within international politics. Indeed, the International Chamber of Commerce (ICC) hosted the meeting on a ship, Statsraad Lehmkuhl, in Bergen, Norway, that led to the creation of the BCSD. The ICC invited Chief Executive Officers (CEOs), diplomats, and other influential business leaders to discuss business and industry's commitment to ensuring continued economic progress while at the same time protecting the environment as a side event at a regional meeting for Europe and North America in advance of UNCED [27]. Attendees at the meeting laid the foundation for the BCSD, the first and most visible pro-business, pro-sustainable development NGO. In retrospect, the organization and insertion of BINGOs into the UN environmental mega-conference system at Rio was among the more significant contributions to global environmental governance that occurred as part of the Earth Summit. The mere presence of groups such as the BCSD fundamentally improved the prospects for regime formation, if for no other reason than serving as a lightning rod for activity. For the BCSD to have credibility within the UN sphere of influence, it would need to deliver more substantial changes in environmental behavior from member organizations. Further, BCSD members would need to avoid corporate "greenwashing," advertising corporate products as environmentally benign when the product damages the environment.

Further, member companies of BCSD opened themselves to more direct criticism from NGOs, particularly those promoting increased environmental protection and supporting development in the South. NGOs, for their part, realized that both businesses and the NGOs that represented them were moving away from their entrenchment on the separation of business from normative values. On the one hand, the more intrepid organizations such as the IIED began to form partnerships with them to explore how this alternative socioeconomic model might work in the future [27]. On the other hand, the more suspicious NGOs wasted no time in declaring the BCSD and their premier documents, *Changing Course* and the *Declaration of the Business Council for Sustainable Development*, as corporate greenwash [19].

The Earth Summit may not have lived up to expectations as far as establishing a new Earth Charter for all countries to follow. Still, it did create a complex framework that significantly expanded the number of environmental issues addressed under the auspices of the UN. Two new major regimes for climate change and biodiversity appear to have been established. Sustainable development gained a foothold as an emerging paradigm within global affairs with the publication of Agenda 21 and the creation of the CSD. Last, Agenda 21 recommended a new treaty, the Convention to Combat Desertification, that opened for signature in 1994.

Equally noteworthy, the Earth Summit opened environmental negotiations to NGO participation. NGO energy did not coalesce around environmental norms but rather splintered around ideological lines including, but not limited to environmental NGOs and business and industry NGOs. No longer would states be able to negotiate treaties in closed corridors. This lasting impact allowed for more voices to contribute to the future of our Planet Earth.

References

1. Adams TL Jr, Martinez-Aragon J (1992) Setting the stage for the Earth Summit: Brazil 1992. Envtl L Rep News Anal 22(3):10190–10200
2. Aksu E (2018) The UN's role in historical context: impact of structural tensions and thresholds. In: Lawler P (ed) The United Nations, intra-state peacekeeping and normative change. Manchester University Press, Manchester, pp 43–75
3. Bernstein J (1992) "A Canadian primer for PrepCom IV." 31 Jan 1992 The Earth Summit NGOs CD-ROM
4. Bernstein et al (1992) PrepCom IV: the final stop on the road to Rio. Network 92:16
5. Borowy I (2013) Defining sustainable development for our common future: a history of the world commission on environment and development. Brundtland Commission Routledge, London

6. Caldwell LK (1995) Necessity for organizational change: implementing the Rio Agenda of 1992. In: Bartlett RV et al (eds) International organizations and environmental policy. Greenwood Press, London, pp 19–35
7. Dimitrov RS (2020) Empty institutions in global environmental politics. Int Stud Rev 22(3):626–650
8. Egelston AE (2012) Sustainable development: a history. Springer, New York
9. Fletcher SR (1992) Earth summit summary. United Nations conference on environment and development (UNCED). Brazil 1992 Washington DC: congressional research service major studies and issue briefs of the congressional reference service supplement; 1993, reel 1, fr. 00082
10. Gallon (1992) The environment liaison centre international a short history. Environment Liaison Centre International, Nairobi
11. Halpern SL (1993) The United Nations conference on environment and development: process and documentation. The Academic Council on the United Nations System Providence
12. Iida K (1998) Third world solidarity: the group of 77 in the UN general assembly. Int Organ 42(2):375–395
13. IUCN (1980) World conservation strategy living resource conservation for sustainable development. IUCN Gland, Switzerland
14. Kütting G (2000) Environment, society, and international relations: towards more effective international environmental agreements. Routledge, New York
15. Kütting G (2004) Globalization and the environment: moving beyond neoliberal institutionalism. Int J Peace Stud 9(1):29–46
16. Lafferty WM, Meadowcroft J (2000) Implementing sustainable developments and initiatives in high consumption societies. Oxford University Press, Oxford
17. Polunin N (1982) UNEP governing council 'session of a special character', held in the Kenyatta international conference centre, Nairobi, Kenya, during 10–18 May 1982. Environmental Conservation 9(2):169–170
18. Rechkemmer A (2003) Rio and the origins of global environmental governance. Sicherh Frieden 21(3/4):173–178
19. Robins N (2003) Profit in need? Business and sustainable development. In: Cross N (ed) Evidence for hope the search for sustainable development. Earthscan London, 204–219
20. Schmidheiny S (1992) Changing course: a global business perspective on development and the environment. MIT Press, Cambridge, Massachusetts
21. Simms A (1993) If not then, when? non-governmental organisations and the earth summit process. Environ Politics 2(1):94–100
22. Srinivas RN (1994) UNCED and the politics of sustainable development. Ph.D. thesis Department of Environmental Studies. State University of New York, College of Environmental Sciences and Forestry
23. Streck C (2001) The global environment facility—a role model for international governance. Global Environ Polit 1(2):71–94
24. Strong MF (2000) Where on earth are we going? Alfred A Knopf Toronto, Canada
25. Struthers D (1983) The United Nations environment programme after decade: the Nairobi session of special character, May 1981 Denver J Int Law Policy 12(3):269–284
26. The Economist (2003) A greener bush. https://www.economist.com/leaders/2003/02/13/a-greener-bush. Accessed 16 Jan 2022
27. Timberlake L (2006) Catalyzing change: a short history of the WBCSD. http://wbcsdservers.org/wbcsdpublications/cd_files/datas/wbcsd/corporate/pdf/CatalyzingChange-A%20ShortHistoryOfTheWBCSD.pdf. Accessed 22 Dec 2020
28. UN (1992) Report of the United Nations conference on environment and development. A/CONF.151/26/Rev.1
29. UN (1989) United Nations conference on environment and development. A/RES/44/228
30. UN (1991) Report of the preparatory committee for the United Nations conference on environment and development. A/45/46

31. UN (1992) Establishment of an intergovernmental negotiating committee for the elaboration of an international convention to combat desertification in those countries experiencing serious drought and/or desertification, particularly in Africa. A/RES/47/188
32. UNCED (1992) In our hands Earth Summit '92 a reference booklet about the UNCED. An earth summit publication: Number 2
33. WCED (1987) Our common future. Oxford University Press, New York
34. Walker RA (2004) Multilateral conferences purposeful international negotiation. Palgrave, New York
35. Willetts P (1996) From Stockholm to Rio and beyond: the impact of the environmental movement on the United Nations consultative arrangements for ngos. Rev Int Stud 22(1):57–80
36. Young OR (2011) Effectiveness of international environmental regimes: existing knowledge, cutting-edge themes, and research strategies. P Natl Acad Sci 108(50):19853–19860

Chapter 8
Climate Change and Global Warming

Abstract With the realization that greenhouse gases increased the average temperature of the Earth's surface, states gathered to prevent climate change. While the negotiation of the United Nations Framework Convention on Climate Change occurred quickly, the Kyoto Protocol proved much more controversial. By 2001, irreconcilable differences in the stringency of targets between the United States and China along with attempts to force the United States to make reductions via industrial energy efficiency led to the collapse of negotiations at COP 6 in The Hague, Netherlands. This unprecedented situation also impacted academic scholarship, as conventional wisdom suggested that the climate change regime had completely formed.

Keywords Climate change · Kyoto protocol · Regime theory · Kyoto mechanisms · Intergovernmental panel on climate change · Carbon credits

After the lofty rhetoric and the inspirational speeches of the Earth Summit at Rio, sustainable development and the international environmental treaties appeared to take center stage at long last on the global agenda. This new socioeconomic paradigm seemed poised to cure the perceived ills associated with capitalism as the dominant economic system, and the peace brought about by the collapse of the Cold War brought hope that attention would shift from traditional military posturing to more sensitive and sophisticated work on development goals for the millions of people locked out of the prosperity of the global North.

One symbol of this new optimism included new voices added to traditional calls from the scientific community and self-proclaimed representatives of the environment. At long last, international pressure on national governments to write a comprehensive treaty to halt climate change became an increasingly popular celebrity cause. Appeals were made to individuals across the globe to become politically active and move governments to commit to stronger actions both within the climate change negotiations and in domestic legislation to avoid future climate change. These appeals came from well-known adult environmental activists like Bill McKibben to new political phenoms, such as teenage climate activist Greta Thunberg. She believes that world leaders' failure to act on this vital issue requires youth to take on the mantle of climate activists. As of mid-July 2020, Greta's School Strike for Climate occurred for the 99th Friday since the first strike in August 2018. These types of student-led events have inspired thousands of people from around the globe to participate in similar events.

Greta's speech to the UN and other world leaders captured the frustration of youth who came of age hearing about the dangers of climate change. And they believed that this single issue, with all its complexities, controversies, and challenges, may well be the most critical policy issue to occur within their lifetimes [21]. Before the COVID-19 pandemic, worry about climate change intensified to the point where a new affliction, climate anxiety, entered the international lexicon to describe the psychological strain associated with this fear [19]. McGinn [17] wrote for Grist magazine that climate anxiety was the biggest pop-culture trend for 2019.

How, then, did the international environmental optimism after the completion of the Montreal Protocol give way to the depression and pessimism of climate change? The two treaties utilize the same framework-protocol structure, deal with interrelated environmental science phenomena, and occur so close together that many of the same individuals were involved, at least at the beginning, of the climate change negotiations. Diplomats who succeeded at Montreal consciously attempted to replicate the successes of ozone diplomacy by applying the lessons learned to climate change.

This chapter seeks to answer these questions by investigating the history of climate change negotiations. Section one reviews the science underlying climate diplomacy, including the work of the Intergovernmental Panel on Climate Change (IPCC). Section two reviews the actions diplomats took to create the framework treaty, beginning with Malta's call to establish the Intergovernmental Negotiating Committee (INC) that started the international diplomatic processes in 1988. Section three discusses the creation of the Kyoto Protocol in 1997. The review of historical events in this chapter ends with the eventual withdrawal of the United States support by then President George W. Bush in 2001. The remainder of the history of climate change after the United States withdrawal is found in Chap. 13. The chapter concludes by reviewing and summarizing scholarly developments, particularly surrounding regime theory and the role of NGO actors through 2001.

8.1 The Science, the Skeptics, and the IPCC

Greenhouse gases refer to those gases that trap heat in the atmosphere by absorbing radiation from the surface of the Earth. Perhaps the most common simple explanation of the effects of the greenhouse effect is to consider the Earth with a weighted blanket covering its entire surface. While that blanket might appear warm and cozy, heat builds up underneath the blanket with disastrous consequences for all life on earth. In a more precise, scientific sense, greenhouse gases, such as carbon dioxide and methane, absorb radiation from the sun and reflect heat toward the Earth's surface. Swedish chemist Arrhenius [1] described this basic greenhouse effect. Many other scientists have also expanded our understanding from this basic climate knowledge. In addition to carbon dioxide and methane, nitrous oxide, CFCs, and HFCs also destroy ozone in the stratosphere (see Chap. 5) and contribute to the greenhouse effect. Thus, there is not a single chemical of concern, but rather multiple chemicals that contribute in differing amounts to the climate problem.

Each of these greenhouse gases stays in the atmosphere for differing lengths of time, meaning that their destructive properties are not identical. Scientists account for this through the creation of the global warming potential that utilizes carbon dioxide as the benchmark with a value of 1. Each chemical's ability to heat the earth is compared to carbon dioxide. For example, the IPCC is a scientific body that reports on environmental science issues to governments. Their Fifth Climate Assessment Report stated that methane has a global warming potential of 28, meaning that 1 ton of methane causes the same heating effect as 28 tons of carbon dioxide [12]. Chemicals may be of concern because of the frequency of their release into the atmosphere or because of the overwhelming amount of radiation trapped by a minor release.

It is believed the buildup of greenhouse gases in the upper atmosphere is closely linked to the industrial revolution that occurred in the mid-nineteenth Century. The Industrial Revolution is the moment in human history where the energy used to power our society changed as coal became the preferred energy source. The use of coal increased dramatically when steam engines entered everyday use in manufacturing and transportation. The rise in climate change generally parallels the increase in annual temperatures since the beginning of the industrial revolution. Scientists have irrefutably documented the increase of carbon dioxide in the atmosphere and the increase in the Earth's global surface temperature.

Two questions then arise; the first is whether the buildup of carbon dioxide falls within the Earth's ability to adjust easily. The second and more concerning question asks when the tipping point is. Climate modelers attempt to predict that moment when the Earth's ability to absorb the increased carbon dioxide in the atmosphere ends and the Earth's climate is fundamentally altered. Climate skeptics tend to believe neither in the Earth's ability to adjust to a new level of carbon dioxide nor that this is their problem to solve. However, the more mainstream view is that action is needed now to ensure the Earth's climate does not experience rapid changes in the concentration of greenhouse gases consequently inflicting severe impacts on all who live on the planet.

Unlike some previous chemicals, preventing greenhouse gases from reaching the atmosphere is not as simple as banning aerosols in cosmetic products as with the ozone layer. Instead, carbon dioxide occurs as a direct result of combustion. In other words, carbon dioxide is released every time a fossil fuel such as oil, coal, or gasoline is consumed. Unfortunately, options for reducing the amount of carbon dioxide reaching the atmosphere tend to be limited as no single "end of pipe" capture system exists for carbon dioxide. Consequently, eliminating greenhouse gases from the atmosphere requires industrial consumers to refine their manufacturing processes to be more energy-efficient or incorporate non-fossil fuel energy sources into industrial processes. Historically, electricity generation and transportation contribute the most to global warming as these two sectors generally consume the most fossil fuel. As energy is required in all manufacturing processes, industrial applications contribute significantly to global warming, as do consumer and residential building spaces.

Individual households, then, have the opportunity to shift their consumption patterns to limit the amount of greenhouse gases released into the atmosphere. Some steps that a household may take to lower their carbon footprint can reduce the energy expended. Such actions may include turning off the lights when leaving a room, limiting the amount of air conditioning and heating used, driving fewer miles in the car, or taking public transportation when available.

Mother Nature also provides help with reducing carbon in the atmosphere through the removal of carbon as part of the carbon cycle. More commonly known as a carbon sink, any manufactured or natural feature that absorbs more carbon than it releases may earn the designation of a carbon sink. Several naturally important carbon sinks include plants, the oceans, and soils. Manmade reservoirs, albeit controversial, include injecting carbon dioxide underground in a technique called carbon capture and storage. This technique captures carbon dioxide from an industrial plant and injects the carbon dioxide underground in a storage reservoir, such as a salt cavern.

A faction of global public citizens, known as climate skeptics, disagree with the mainstream view that climate change has a manmade component. Writing in 1998, after the negotiations concluded for the Kyoto Protocol, Jacoby et al. [13] pointed out that predicting future climate relied on a series of unknowns, including portions of the carbon cycle, particularly the role of oceans. This group of authors also noted that climate models rely on estimates of future greenhouse gas usage, including predictions of population growth and technological adaptation, two areas notoriously difficult to forecast accurately.

One of the personal questions that must be answered then is whether the risk of catastrophic climate change justifies the sacrifices of many individuals. There are many various ways to answer this question. However, many of our most knowledgeable scientists judge the benefits of avoiding climate change to be well worth the individual sacrifices, including their own. Jacoby et al. [13] questions the failure to act in light of these risks.

Given the variety of beliefs about the causes, risks, and importance of climate change, it is perhaps unsurprising that public support for climate change does not remain constant, public support changes over time. Environmentalists' concern with climate skepticism stems from the realization that global public pressure on diplomats to negotiate strong binding treaties fades as more people express skepticism. However, it is important to note for this chapter that climate change never enjoyed widespread public support compared to public support for closing the ozone hole. In the case of the ozone hole, public support successfully lobbied governments to act. While there have been many successful attempts to lobby for climate change action, they have not occurred worldwide at the same time.

As with many of the other international environmental protocols, the early history of this issue area begins with the scientists. The First World Climate Conference occurred in 1979 when scientists examined the impact of human activity on the climate. The meeting concluded with the idea to create a World Climate Research Program under the auspices of the WMO, the UNEP, and the International Council of Science. In addition to the First World Climate Conference, other important meetings that evaluated the current status of climate science included the Villach meeting in October 1987, the Bellagio conference in November 1987, and the Toronto Conference in June 1988. During the Toronto Conference, then Prime Minister Margaret Thatcher of the United Kingdom called for international action to address climate change. Thatcher's presence appeared to signal a shift in favor of support for climate change. The "Iron Lady," as Prime Minister Thatcher had been dubbed, was known for her conservative, unshakeable approach to economics. Consequently, environmentalists expected her to oppose climate change on economic grounds.

Later that year, the UNEP and the WMO formalized the creation of the IPCC. The group produces period assessments on climate change science and makes recommendations about mitigating, adapting, and creating new policies to combat climate change for national governments. Thus, the IPCC does not produce new research but rather synthesizes research published through other venues. In addition to these assessment reports, the IPCC also publishes special reports on specific topics upon request of a member government or international organization.

One of the key documents that the IPCC utilizes to communicate with other actors is the Assessment Reports. While 195 governments formally make up the membership of the IPCC, the reality is that many hundreds of scientists participate in the process of creating an Assessment Report. The reports are intended to be scientifically and politically relevant, yet nevertheless, avoid giving policy advice; the authors intend for the reports to be politically neutral. The IPCC issued the First Assessment Report in 1990, concluding that human activities contribute substantially to climate change. This timely report assisted in elevating the status of the issue as it occurred in the international negotiating committee for climate change.

The IPCC operated three working groups—I, II, and III in its original structure. Working Group I focused on assessing the science, while Working Group II focused on impacts, adaptation, and mitigation. Working Group III concentrated on economic and social dimensions related to climate change. In essence, Working Group I focused almost exclusively on science, while Working Groups II and III focused on policy impacts, an area where scientists may be more cautious [16].

8.2 The UNFCCC

Calls for a formal treaty to limit the amount of greenhouse gases occurred sporadically from scientists and the environmental community throughout the 1980s. One of the more widely spread calls for action on climate change occurred as part of the Brundtland report [20]. Malta's call for an international treaty on climate change in 1988 was formally accepted by the UN GA. As a result of this shift in sponsoring bodies from the UNEP to the UN GA, many more states became involved in the process of creating what became the UNFCCC. At once, both complicating the process of creating a treaty but also simultaneously raising hope regarding the strength of the international community's collective will to combat climate change as more countries engaged in the negotiating process. The change of sponsorship to the UN GA also signified a shift toward including developmental concerns within international treaties. The G77 and China voting bloc numerically dominates the UN GA, thus raising the importance of the Southern negotiating agenda within this issue area as the global South controls the UN GA [16].

In the aftermath of the success of the Montreal Protocol, the framework-convention recipe became the immediate blueprint for the climate change treaty. This outcome is not surprising given that the two treaties occur close together in the overall trajectory of international environmental diplomatic history. Additionally, many of the same ambassadors and technical advisors work on both issues [14]. Overall, the UNFCCC stated an overall goal of stabilizing greenhouse gas emissions at the 1990 emissions level to prevent a catastrophic alteration of the chemical balance of the upper atmosphere. It is important to note that this goal does allow for some gradual climate change as long as adaptation and mitigation of consequences occurs. The 1990 baseline selection process was a political decision meant to convey urgency rather than a scientific one. Regardless, very few countries had accurate data, if a country had data at all.

The UNFCCC also contains general principles, basic obligations of states, and procedural provisions for conducting future conversations. Gupta [8] identified five principles within the climate change regime; three of these principles tend to reoccur from one treaty to another. First, diplomats opted to utilize the precautionary principle. Second, equity considerations between the North (Annex I countries in the context of this issue area) and the South (also known as Annex II countries) should also be taken into consideration using the concept of common but differentiated responsibilities. This concept points out that while everyone will suffer from the negative impacts of climate change, poverty-stricken countries have neither the resources nor the technological capabilities to implement potential solutions. Third, diplomats included sustainable development within the treaty. Indeed, given the intertwined history of the Rio Earth Summit and the climate change treaty, the two terms are sometimes used as synonyms, albeit inappropriately.

Within the Framework Convention on Climate Change, all national governments agree to create and share national greenhouse gas emissions inventory, including greenhouse gas emissions and removals by greenhouse gas sinks. National governments should also cooperate on creating and implementing climate policy, including, but not limited to, sharing scientific knowledge, creating and developing new control technologies, and improving education about related scientific and technological fields.

Procedurally, the UNFCCC creates the COP consisting of all signatories of the Convention that meets regularly. Two subsidiary bodies with open membership, the Subsidiary Body for Scientific and Technological Advice and the Subsidiary Body for Implementation, advise the COP as climate change remains a complex issue. A Secretariat, located in Bonn, Germany, also provides logistical assistance and administrative staff. Last, the IPCC meets to assess the state of the art of climate science and informs all governments about potential adaptation and mitigation strategies.

While the UNFCCC specified an overall goal for the negotiations and a process to continue conversations, it did not specify binding emission targets for individual nations. This rather important detail was left for future discussions upon ratification.

Negotiations on this all-important detail began with the COP 1, held in 1995 in Berlin, Germany. The national governments meeting at COP 1 produced the so-called Berlin Mandate, an agreement that the voluntary commitments pledged in the UNFCCC should give way to mandatory commitments and establish a time frame to negotiate the magnitude of the reductions.

The IPCC's Working Group I helped to strengthen the political resolve to add to the UNFCCC when this group in 1995 wrote that there was a very low probability that climate change was occurring through natural variability only [10]. This assessment heralded a change in the understanding of the causes of climate change. It also heightened concern about the future of the planet. This assessment documented the falseness of the belief of many individuals, namely that humankind was not able to impact the global climate.

8.3 The Kyoto Protocol

This two-year process culminated in 1997 with the Kyoto Protocol negotiated at COP 3. Oberthür and Ott [18] provided an extensive review of the negotiating history and content of the Kyoto Protocol. The Kyoto Protocol divides countries into Annex I countries with binding emission targets and Annex II countries without binding emission targets. Annex I countries typically include signatories to the North Atlantic Treaty Organization or the Warsaw Pact (the North), while Annex II countries consist of the global South and China.

The Kyoto Protocol consists of a series of carefully crafted compromises between multiple divergent approaches. Significant differences occurred on the topics of what countries should make legally binding commitments, how strong the commitments should be, the use of policies and measures or market mechanisms, and the role of sinks in meeting the overall emission reduction targets.

Overall, the Kyoto Protocol seeks a 5% reduction from 1990 levels from those countries in the North as specified in Annex B for four gases (carbon dioxide, methane, nitrous oxide, and sulfur hexafluoride) and two groups of fluorinated gases in the 2008–2012 time frame. Each country within the North established its own separate legally binding target. For example, the United States agreed to a 7% reduction, while the European Community (now the EU) agreed to an 8% reduction. Countries that were once a part of the Soviet Union also took on legally binding targets; however, the partial economic collapse of communism in these countries meant that these reductions had already occurred. Instead, these countries looked to stimulate economic growth without exceeding their Kyoto target.

The so-called "EU bubble" remains of particular import within the individual countries' reduction commitments. This concession to the EU created individual commitments for each of the EU member states but allowed the group of states to meet their obligations collectively. In other words, if the United Kingdom met its national target, but Spain did not, rather than looking at each country individually, the analysis of whether Spain met its commitments would occur at the supranational level, not the individual level. Jacoby et al. [13] reported that the United States supported the EU bubble concept to lower European resistance to other aspects of the Kyoto Protocol, namely emissions trading.

The roughly 130 countries in the global South did not consent to binding emission targets. While this position makes rational sense from the standpoint of common but differentiated responsibilities, it nevertheless puts the environment at risk for greater climate change impacts. All the oil-rich countries in the Middle East avoided mandatory actions. More problematically for the environment and the future of the climate change negotiations, China, now the world's leader in absolute greenhouse gas emissions, avoided binding targets.

At the heart of the Kyoto Protocol was a carefully crafted compromise between the EU's preference for harmonized policies and measures and the United States' vociferous insistence on market mechanisms, including carbon offsets and emission trading mechanisms, two very different approaches to climate change reductions. Harmonized policies and measures relied on regulatory schemes such as permits and mandatory equipment specifications to reduce allowable emissions over time. In some instances, the harmonized policy and measure phrase implied either an energy tax or a carbon tax. Countries would have certainty that all emission sources would make emission reductions in tandem and that national governments would remain fully in control of the reductions' timing and speed. However, governments run the risk that impacted entities would refuse to adopt new technology quickly due to the complex regulatory scheme. Permitting costs corporations labor and financial resources to implement; corporations can and will delay equipment upgrades to postpone spending additional funding on permitting, even if it increases their tax burden.

Alternatively, emission trading mechanisms require a governmental agency to establish a trading currency representing a fixed amount of pollution allowed to enter the atmosphere. The governmental agency also determines the methodology corporations use to earn credits and monitor actual emissions. Each regulated entity must hold enough allowances to match their actual emissions over a predetermined amount of time. Corporations are then allowed to make reductions in their facilities or purchase emission allowances from other sources, as long as no other environmental laws or regulations are broken in the process of lowering emissions. Unlike harmonized policies and measures, cap and trade mechanisms meant that governments give up control of where precisely the reduction occurs to the regulated entity. Permitting delays may or may not be an issue depending on where the reduction occurred. In other words, corporations may choose to make upgrades in many minor sources to avoid making reductions at larger sources. The environment is unlikely to distinguish between the multiple pathways for reductions due to climate change unless the chemical simultaneously created another negative impact. For example, methane emitted to the atmosphere is frequently found with other malodorous gases (hydrogen sulfide); people living near a methane source may well have a strong preference for this stream to be captured to eliminate the odor.

The Kyoto Protocol effectively created three different emission trading mechanisms—carbon cap and trade allowances, Joint Implementation projects, and Clean

Development Mechanism projects. Each of the three mechanisms interacts with and is entirely interchangeable as far as meeting a countries reduction commitment during the 2008–2012 time period. Countries that exceed their emission reductions goals would be allowed to "bank" their reductions for a future compliance period. However, the Kyoto Protocol did not specify additional time frames for mandatory binding commitments.

Carbon allowances were allocated by governments to individual companies and facilities located within their country. Each country decides what industries to include and how to allocate allowances to these industries. While many scholars and environmentalists advocated allowing industries to purchase the initial allocation at a fixed price (effectively turning the cap and trade mechanism into a carbon tax), many governments gave away the initial allocation for free. National governments also established a registry for tracking emissions and carbon allowances. Companies then had the flexibility to buy or sell allowances based on their production needs and control strategies. Carbon pricing thus fluctuates over time.

Joint Implementation projects allow Annex I countries to earn credits from projects implemented in another country by trading their "assigned amounts." In other words, under the rules of the Kyoto Protocol, a country may sell its excess emissions to another country or another legal entity located in another country. Joint Implementation projects would need approval from both the host and receiver countries. Last, the EU inserted a phrase that emission reductions should be "supplemental" to domestic activities. In other words, countries should not be able to utilize joint implementation projects to avoid making reductions at home.

When combined with the economic decline of the former-Soviet states, the possibility of Joint Implementation meant that most, if not all, of the Soviet bloc would have extra emissions available to sell. While this economic windfall would undoubtedly help the former Soviet Union states recover from the end of communism from an environmental standpoint, it created the potential for countries such as the United States or the EU to buy excess allowances without making any operational changes through the end of the first commitment period.

The Clean Development Mechanisms allowed carbon reduction projects that created real (not only on paper) carbon emissions reductions located in Annex II countries that did not have a binding commitment to equal access to the emission trading mechanisms. Consequently, Southern countries discovered a new revenue stream that could bring new finances and technologies into their territory. By COP 5 in Bonn, Germany, a scant three years later, Southern countries would insist upon and quickly receive a "prompt start" to the Clean Development Mechanism that allowed projects to earn credits before the beginning of the 2008–2012 time period, giving these projects a political and economic advantage.

Last, the Kyoto Protocol contains the promise of "new and additional" financial resources for developing countries to industrialize their economies while avoiding the environmental damages associated with industrialization. The developing countries interpretation of this portion of the treaty includes the notion that developed countries will pay for "the agreed full, incremental costs of developing countries," including technology transfer and all resources needed to fulfill any compliance, adaptation, or mitigation consequences from climate change [8]. Consequently, the climate change regime may be reasonably interpreted as both an environmental and developmental treaty.

Ratification of the Kyoto Protocol occurred when 55 countries representing at least 55% of the total greenhouse gas emissions inventory covered by the binding commitments agreed to the Protocol. Thus, the United States became a key figure in ratification as it accounts for 36% of the emissions inventory at the time. While the United States may not unilaterally block ratification, the absence of the United States would signal a key political weakness.

While then Vice-President Al Gore attended the Kyoto conference and participated in the negotiating process, ratifying a treaty of this magnitude within the United States is not straightforward. The United States operates with independent control of the legislative and executive branches of government; the executive branch has the authority to negotiate and sign a treaty, but the legislative branch, in this case, the United States Senate, has sole authority to ratify and approve a treaty. Treaties signed by the executive branch but not submitted to the Senate for ratification are not legally binding upon the United States.

The Byrd–Hagel resolution of 1997 [4] expressed the collective thought of the United States Senate. Senators voted 95–0 on a non-binding resolution that stated that the United States Senate would not ratify any climate change treaty that did not contain binding commitments for developing countries. One way to look at the United States' behavior in this instance is that the executive branch negotiated a treaty that the legislative branch would not accept. This clear conflict between the legislative and the executive branches typically does not occur in other countries where the executive branch always aligns with the legislative branch to control the government itself politically. This nuance of American foreign policy historically frustrates other countries because of the absence of a guarantee that the Senate would formally ratify the treaty.

After the euphoric high of agreeing to the Kyoto Protocol at the proverbial "eleventh hour" in Kyoto, Japan, the signatories to this document needed to complete the negotiations that would allow for the treaty's implementation, in particular the negotiations of the policies and measures as well as the details for implementation of the so-called Kyoto Mechanisms. COP 4, meeting in Buenos Aires, Argentina, began this arduous task. The most noticeable outcome from COP 4 undoubtedly was the creation of the Buenos Aires Plan of Action, which laid out a timeline for flushing out the details omitted from the Kyoto Protocol. Diplomats agreed to finalize the operationalization of the Kyoto Mechanisms by COP 6, tentatively scheduled for late 2000.

The political impetus for adhering to this timeline undoubtedly benefitted from the United States' announcement of their signing of the Kyoto Protocol. ENB [6] noted that this development may be one of the more important outcomes during COP 4 under the guidance of President Clinton. However, the Buenos Aires meeting also foreshadowed the difficulties of the climate change arena along two dimensions. The first dimension involved the commitments, or lack thereof, from developing countries. While Argentina announced its intent to take on a voluntary commitment, the backlash from the announcement hardened resistance from other Southern countries, most notably China, who feared international pressure to limit its economic growth. The second dimension came from the process itself.

One particularly welcome change in the 1999–2000 time frame involved the Southern countries increasing interest in the Clean Development Mechanism. While the developing countries were initially skeptical about the Kyoto mechanisms, the realization that the Clean Development Mechanism incentivized investments in clean technology changed their negotiating stance on this issue. Some experts [13] believed that the lowest cost greenhouse gas reductions will occur in Southern states. These states will then be able to sell their Clean Development Mechanism credits for a profit. Consequently, the overall value of the Clean Development Mechanism program may serve as an acceptable mechanism for a wealth transfer from North to South.

That being said, global angst over the entire emission trading scheme resulted in significant discussion around the issues of "fungibility," whether a carbon credit created under one of the Kyoto Mechanisms could be used within a different mechanism, and around "additionality," a yet to be defined term that represented the extent to which "business as usual" would be allowed under the Kyoto Protocol. Of the two terms, additionality would prove to be the more difficult to resolve. The G-77 & China understood that anything less than full fungibility hurt their potential future revenue sources and aligned with the United States, which wanted cheap carbon credits. COP 6 in The Hague, Netherlands proved to be the pivotal moment for the success, or lack thereof, for the Kyoto Protocol. The disagreements between the United States and the European Union about the operationalization of emission trading proved too high a hurdle to overcome. While the discussions about fungibility were mostly amicable and resolved in favor of complete substitution, additionality remained unresolved.

However, diplomats' inability to resolve conflict about the role of "carbon sinks" sank the negotiating session. The United States, in particular, contains multiple forests that soak carbon from the atmosphere, and it wanted to count these reductions toward its compliance obligations. Potentially, industries in the United States could avoid reducing industrial emissions by planting more trees if reductions from natural sinks were counted. Other countries such as the EU do not have this same potential; thus, EU-based industries could face more operating costs, placing their products at an international trade disadvantage. Additionally, the EU preferred emission reductions from industrial sources over other potential reduction mechanisms and sought to limit the use of sinks within emission trading schemes.

The problem was undoubtedly compounded by the uncertainty surrounding the 2000 presidential election. COP 6 coincided with Florida's attempted recount of ballots cast in the presidential election. Jacoby and Reiner [14] suggested that Vice President Gore's absence from the negotiating session doomed the diplomatic negotiations. Certainly, many environmentalists in the immediate aftermath of the discovery that the Democratic party lost control of the American presidency bemoaned the lost opportunity to come to an agreement as they expected incoming President George W. Bush would be much more insistent about the inclusion of sinks toward meeting the US's Kyoto commitments [11].

8.4   To Regime or Not to Regime

In the immediate aftermath of the Kyoto Protocol, it certainly appeared as if the international negotiators had successfully replicated the successes of the Vienna Convention and the Montreal Protocol for substances that deplete the ozone layer. Jacoby et al. [13] divided the analysis of the impact of the Kyoto Protocol into three groups, the optimists, the pessimists, and the skeptics. Each of these perspectives offers insight into the climate change regime. The optimists seemed confident that the climate regime would successfully adopt the norms represented within the Kyoto Protocol, including additionality, the precautionary principle, and common but differentiated responsibilities. National governments had committed to start the journey toward decoupling carbon emissions from energy use. The political will to protect the environment would prod naysayers into more environmentally benign actions. If individual corporations could not be swayed into doing the right thing (reduce emissions) for the right reasons (adopt new norms), then the financial benefits and potential lost revenue would nevertheless win out. The pessimists believed that the developed countries traded future economic success to avoid diplomatic embarrassment. This group typically believes that the financial cost of emission controls for greenhouse gases should not be paid until scientists can determine the seriousness of future climate change impacts. The skeptics fell somewhere in the middle of these two extremes. On the one hand, the skeptics believe that climate change is occurring and that meaningful commitment to lowering the concentration of greenhouse gases in the atmosphere would benefit both people and the planet. However, the obligations contained in the Kyoto Protocol, or more correctly, the Berlin Mandate, commit national governments to short-term commitments to reduce greenhouse gas emissions that will not result in lower atmospheric concentrations of greenhouse gases.

In more academic phrasing, scholars expected that national governments had successfully formed a new international regime [3]. Scholars looking solely at the wording of the Kyoto Protocol certainly had reason to believe that national governments articulated a clear set of norms, values, beliefs, and decision-making procedures. However, the concern with regimes revolves around the last phrase of [15] definition where actor's behavior should be predictable and in line with the treaty's requirements [6]. This phrase implies that states undertook some type of implementation activity and states' non-compliance with the treaty comes with consequences, even if that punishment is mainly symbolic.

In this respect, it is relatively clear that scholarly publications that heralded the rise of a strong climate change regime were premature. By 2001, the realities of the situation looked rather bleak and depressing. Without the United States and its considerable economic and political leadership, the ability of the other countries to make meaningful reductions and keep the promises of the Kyoto Protocol quickly disappeared. Despite the immediate attempt of the EU to politically isolate the United States, the world's remaining superpower proved too strong to humiliate into making the mandatory reductions negotiated at Kyoto, Japan. The option for the United States to save face by adhering to the Protocol's reductions without ratifying the treaty slowly ebbed away as the time for the first commitment period approached. Scholarly certainty about regime formation disappeared as a new question emerged—to what extent can a regime form without the world's sole superpower?

In the absence of formal negotiating processes, environmentalists pinned their hopes on the parallel informal structures outside states' control. Recalling that the role of science within international environmental affairs continues to explore the intricacies of global politics, the climate change regime also offers insights into this relationship. Haas' [9] groundbreaking research on epistemic communities typically withstands close scrutiny on many different environmental issues. The consensus among scholars is that an epistemic community formed for climate change as well [7]. Further, Lanchbery and Victor [16] analyzed the relationship between the IPCC and the INC for the UNFCCC in the early 1990s. They conclude that the IPCC's effectiveness is most strongly linked to its function; that the IPCC performs admirably when asked about purely scientific matters. However, Lanchbery and Victor also concluded that the IPCC's performance languished when asked about future impacts that are inherently unknown or about response strategies and policy options. One lesson learned from the climate regime may be that scientists prefer to focus on science rather than drafting policy proposals. However, the fact that this epistemic community remained intact meant that scientific research could be used in the future to pressure the United States to take action.

While much of the research on climate change revolves around the actions and preferences of states, or alternatively, the role of science and scientists in determining the extent of the problem and potential policy solutions for these same states, scholarly research also branched out to investigate the role of non-state actors, especially those of NGOs. After NGOs' successes in the ozone regime and taking into account the emphasis on NGO participation in the Earth Summit, NGOs formed an essential relationship with the UN and with each other. That is not to say that NGOs experienced a breakthrough moment. They did not. It is much more correct to say that the acceptance of NGOs within international environmental diplomacy occurred one conversation at a time. NGOs accomplished diplomatic acceptance by proving their technical expertise. They also gained respect due to the inherent value of reflecting global public opinion into the various negotiating processes.

It is also patently unfair to say that all NGOs reflected unanimity of thought or desires. Indeed, it has become something of a tradition within academic circles to carve out business and industry groups from the remainder of the non-governmental community and, for convenience's sake, that same technique, although not grounded within reality, will serve as a good starting point for a review of the activities of NGOs here. A simplistic analysis of the split within the NGOs divides the community into two pieces—environmental NGOs such as Greenpeace or Climate Network Europe stand in opposition to those business and industry groups such as the International Chamber of Commerce or the Global Climate Coalition. Environmental groups are assumed to support environmental treaties representing more stringent environmental protection while business and industry groups are presumed to oppose both the treaties and the increased environmental quality.

Gough and Shackley [7] assessed the role of NGOs and epistemic communities. They concluded that NGOs moved from outsiders who organized public awareness campaigns to influence outcomes on a single-issue area to insiders who actively partnered to develop negotiating principles. Betsill [2] provided more details by noting that NGOs influenced the treaty text itself, a remarkable achievement for groups that have a history of being told to "go home."

Long after the mirage of a unified climate change regime disappeared, Chasek and Wagner [5] observed that the emerging principle of universal participation in international environmental negotiations may well have negatively impacted the likelihood of producing effective treaties. As a fixed number of sovereign states and an increasing number of NGOs sought to influence treaty outcomes, the more difficult achieving consensus or unanimity becomes. From the highs of the Kyoto Protocol in 1997, to the lows of the United States withdrawal from the Kyoto Protocol in 2001, scientists and proponents of a strict climate change regime remain convinced that atmospheric warming caused by the addition of greenhouse gases in the upper atmosphere is the defining issue of this generation, and perhaps, of the generation to come.

References

1. Arrhenius S (1896) XXXI On the influence of carbonic acid in the air upon the temperature of the ground. London Edinb Dublin Philos 41(251):237–276
2. Betsill MM (2000) Greens in the greenhouse: environmental NGO's, international norms and the politics of global climate change. Dissertation, University of Colorado
3. Bodansky D (2001) The history of the climate change regime. In: Luterbacher U, Sprinz DF (eds) International relations and global climate change. MIT Press, Cambridge, Massachusetts
4. Byrd-Hagel Congressional Record October 3, 1997 (senate) S10308–S10311
5. Chasek P, Wagner LM (2016) Breaking the mold: a new type of multilateral sustainable development negotiation. Int Environ Agreem 16:397–413
6. Depledge J (2013) The organization of global negotiations: constructing the climate change regime. Routledge, London
7. ENB (1998) Report of the fourth conference of the parties to the UN framework convention on climate change: 2–13 Nov 1998. 12(97):1–14
8. Gough C, Shackley S (2001) The respectable politics of climate change: the epistemic communities and NGOs. Int Aff 77(2):329–346
9. Gupta J (2010) A history of international climate change policy. Wires Clim Change 1(5):636–653
10. Haas PM (1990) Saving the mediterranean: the politics of international environmental cooperation. Columbia University Press, New York
11. Houghton JT et al (1995) Climate change 1995: the science of climate change contribution of working group I to the second assessment report of the intergovernmental panel on climate change. Cambridge University Press, Cambridge
12. Houlder V (2000) Greenhouse gases environmental campaigners call for talks to resume: EU and US under pressure on climate deal. Financial Times. December 2, p 6
13. IPCC (2014) Climate change 2014: synthesis report. contribution of working groups I, II and III to the fifth assessment report of the intergovernmental panel on climate change. IPCC, Geneva
14. Jacoby HD et al (1998) Kyoto's unfinished business. Foreign Aff 77(4):54–66
15. Jacoby HD, Reiner DM (2001) Getting climate policy on track after The Hague. Int Aff 77(2):297–312

16. Krasner SD (1982) Structural causes and regime consequences: regimes as intervening variables. Int Org 36(2):185–205
17. Lanchbery J, Victor D (1995) The role of science in the global climate negotiations. In: Bergesen HO, Parmann G (eds) Green globe yearbook of international co-operation on environment and development 1996: an independent publication on environment and development from the Fridtjof Nansen Institute Norway. Oxford University Press Oxford, pp 29–39
18. McGinn M (2019) 2019's biggest pop-culture trend was climate anxiety. Grist https://grist.org/politics/2019s-biggest-pop-culture-trend-was-climate-anxiety/. Accessed 16 Jan 2022
19. Oberthür S, Ott HE (1999) The Kyoto protocol international climate policy for the 21st century. Springer, New York
20. Pihkala P (2019) Climate anxiety. MIELI mental health Finland. Helsinki
21. WCED (1987) Our common future. Oxford University Press, New York
22. Weise E (2019) 'How dare you?' Read Greta Thunberg's emotional climate change speech to the UN and World leaders. USA Today https://www.usatoday.com/story/news/2019/09/23/greta-thunberg-tells-un-summit-youth-not-forgive-climate-inaction/2421335001/. Accessed 31 Jan 2022

Chapter 9
Conserving Biodiversity

Abstract Conversations about conserving the natural environment predate the beginning of international negotiations to create a treaty to preserve genetic resources. Early efforts to promote individual species and their habitats through debt-for-nature swaps gave way to an emphasis on protecting genetic resources intact in the ecosystem. As many of the highest diversity sites are in developing countries, these states required additional resources to adequately protect these locations. Thus, states negotiated the Convention on Biological Diversity to assist with creating a regulatory scheme to manage risks to biodiversity, including natural resource depletion, and biotechnologies that might out compete natural organisms. The Convention on Biological Diversity reflects characteristics of both an environmental treaty and a trade agreement. Disagreements about the relationship between the UN sponsored Convention on Biological Diversity and the World Trade Organization's Agreement on Trade-Related Aspects of Intellectual Property Rights dominated scholarship contemporary to these negotiations.

Keywords Convention on Biological Diversity · Cartagena Protocol on Biosafety to the Convention on Biological Diversity · Nagoya Protocol on Access to Genetic Resources and the Fair and Equitable Sharing of Benefits Arising from their Utilization to the Convention on Biological Diversity · Agreement on trade-related aspects of intellectual property rights · World Trade Organization · Genetically modified organism

Deep within the Amazon jungle, indigenous peoples live in harmony with the world's largest tropical rainforest. The mighty Amazon has, for centuries, captured the imagination of global citizens with its impressive tree stands, indigenous peoples, and colorful wildlife. Frequently pictured as an idyllic setting for majestic tropical birds such as the toucan and macaw, the Amazon captured the global public imagination as an area rich in biodiversity and natural beauty. However, the mystique aura generated by the Amazon does not spare this ecosystem from environmental controversy.

Often referred to as the Earth's lungs, the Amazon rainforest stretches across South America, encompassing 2.3 million square miles of canopy cover. The Amazon River basin contains an area significantly larger than the forest proper, with a total size of 3–3.2 million square miles [3]. Approximately 60% of the Amazon lies in the state of Brazil, with Peru controlling 13% and Colombia controlling an additional 10%. The Amazon also stretches into Bolivia, Ecuador, Venezuela, Guyana, and Suriname. Additionally, French Guiana, an overseas department of France, also controls a minor portion of the Amazon. The Amazon River itself is noteworthy as the second-longest river system in the world, behind only the Nile River in Africa. Various attempts have been made to define the headwaters for the Amazon in the Peruvian Andes mountains. Contos and Tripcevich [4] placed the headwaters for the Amazon River at Cordillera Rumi Cruz, Peru.

The Amazon plays a unique role within the climate system for the planet as the rainforest contributes to the hydrological cycle. Transpiration, the release of water as a product of photosynthesis, creates additional moisture that spreads into the air and radiates out from the Amazon. This moisture falls back to the surface as rain and allows crops to exist in the area. Additionally, the Amazon also plays a significant role within the carbon cycle. Photosynthesis, the process by which a plant takes in sunlight, water, and carbon dioxide to create oxygen and energy, makes the Amazon rainforest invaluable in regulating the carbon cycle.

Once a significant source of carbon absorption, scientists now believe that the deforestation of the Amazon contributes to climate change. This burning not only releases the carbon stored inside trees but also decreases the rate of future carbon adsorption through photosynthesis. Moran [23] estimates that significant deforestation of the Amazon occurred beginning in 1975. Silva Junior et al. [38] reported that the Amazon's deforestation rate exceeded its allowable rate by 182% over its target of limiting the destruction to 3925 km^2 in 2020 in Brazil.

The destruction of the Amazon deprives the planet of a necessary defense against climate change and destroys essential habitats for flora and fauna, lowering biological diversity. Environmental groups routinely claim that 10% of the known species on the planet could be living in the Amazon biome [55]. Researchers have not ascertained the properties of this biodiversity, and the Amazon could likely contain genetic materials and naturally occurring chemicals that could spur new technological inventions, including new pharmaceutical medicines.

Causes for deforestation within the Amazon vary according to country. Brazil, for many years, sought to move inland off the coast and pursued a policy of deforestation to build roadways [23]. Economic interests, frequently supported by governmental policies, continue to burn the forest to clear the land for cattle ranching and other agricultural crops [9]. As the deforestation of the Amazon occurs, Brazilians come closer into contact with the indigenous peoples of the Amazon, causing conflict between the civilizations. Indigenous peoples remain highly dependent on an intact Amazon to support their way of life as they tend to eschew modern conveniences and lifestyles. This lifestyle may be threatened by loggers seeking access to the timber within the Amazon who may not respect limits imposed by the government [27].

Given the Amazon's role as the Earth's lungs, other countries have taken note of the status of the Amazon. International pressure on Brazil manifested in the mid-1980s as environmental organizations highlighted the increased deforestation. Conservation biologists utilized their professional prestige to advocate for an end to degrading the tropical rainforest. Environmental NGOs also expressed concern about biodiversity loss. These groups attempted various schemes to protect the Amazon over the last 35 years, all with little lasting impact as the Amazon rainforest shrinks.

This chapter reviews the creation of the CBD and its additional protocols intended to help protect fragile ecosystems like that of the Amazon. The first section presents the rationale and an abbreviated negotiating history for the CBD. The second section reviews the establishment of the CBD at the Rio Earth Summit. Section three describes the Cartagena Protocol on Biosafety to the Convention on Biological Diversity (Cartagena Protocol). Section four discusses the Nagoya Protocol on Access to Genetic Resources and the Fair and Equitable Sharing of Benefits Arising from their Utilization to the Convention on Biological Diversity (Nagoya Protocol). Section five concludes the chapter by discussing scholarly research associated with this treaty.

9.1 Rationale for the Biodiversity Convention

Conversations about conserving the natural environment predate the beginning of international negotiations to create a treaty to preserve genetic resources. In 1980, IUCN, in cooperation with WWF and UNEP, published the World Conservation Strategy, a blueprint for governmental actions to preserve species, habitats, and ecosystems at the national and international levels. The report began with a subtle but important normative shift: conservation and development must be considered in tandem as the need to develop consumed the natural resources that conservation strategies attempt to preserve. Ultimately, the World Conservation Strategy presented conservation priorities and strategy with two caveats: that generalizations necessary to apply globally by necessity neglect local circumstances and that balancing the need for environmental protection with continuing development imperatives proved difficult [12].

In framing sustainable development in this manner, IUCN pointed toward three critical conceptualizations that would impact the creation of the CBD. First, natural resources as part of the Earth's carrying capacity will not last indefinitely. The Earth's carrying capacity represents the idea that natural resources are finite and limited. Once the environmental resources are consumed, the materials are unlikely to be replaced. Further, population growth increases stress on natural resources as individuals strive to meet their needs and increase their wealth. Second, there is an inherent conflict between local and global uses of resources. In many cases, there is intense pressure to consume resources to meet the immediate needs of the local community. However, destroying the local rainforest impacts local communities and triggers global changes. Loggers, farmers, and ranchers that move into the area may engage in deforestation, which negatively impacted the Indigenous Peoples native to the area. Thus, consumption of natural resources directly benefits one group while increasing the likelihood of environmental harm for all. Third, redistributing the benefits of resource conservation enhances the possibility of protecting the resource.

As one group loses more than others by keeping the ecosystem intact, a compensation scheme may be necessary to provide an incentive for avoided consumption of the ecosystem.

McCormick [20] pointed out that IUCN, along with other environmental NGOs, shifted their position on the relationships between environment and development closer to the Southern counties' framing that inequitable distribution of economic wealth contributed to increased consumption of natural resources, including deforestation. Attempts to integrate developing countries into the capitalist economic system through structural adjustment programs advocated by the World Bank and others exacerbated the pre-existing inequalities. The World Bank included these structural adjustment programs as part of the conditions for governments receiving a loan. The program forced countries to implement policies that frequently included lowering government spending and raising taxes while simultaneously opening their domestic markets to international competition. Consequently, developing countries were left in the untenable position of repaying costly loans without ever achieving the desired outcome, that of improving their ability to generate wealth in the global economy.

Aware of the threat to habitats around the globe necessitated by increasingly dire financial circumstances, Lovejoy [19] proposed utilizing debt-for-nature swaps in a New York Times article that received immediate positive interest. Lovejoy, then Vice President of Science for the WWF, shared the concerns of many other conservation biologists; that the debt crisis of the 1980s would force developing countries rich in biodiversity to end spending to protect these ecosystems. Worse yet, the need to raise money to repay the loans would increase pressure to develop the land for agriculture or industrial uses that would permanently alter the landscape.

To begin a debt-for-nature swap, conservation groups needed to acquire the debt obligations from the banks holding the note. Conservation groups raised money for purchasing the debt. The primary mechanism for doing so involved a secondary market that resold loans at a reduced value as the countries' circumstances increased the likelihood of default or actual default had occurred. In rare cases, the bank donated the debt in exchange for publicity. The conservation group then swapped the debt back to the borrower nation in exchange for an agreement with the national government to support environmental projects. In the process, all three groups benefitted. The banks removed the debt from their ledgers; the conservation group achieved its preferred policy outcome, and national governments improved their financial health by retiring debt. While these debt-for-nature swaps occurred in small amounts in comparison to the countries' indebtedness, they nevertheless changed local circumstances in specific locales needing of protection. Additionally, debt-for-nature swaps stimulated international publicity in favor of increased environmental protection.

The totality of circumstances in the mid-1980s created an opportunity to collaborate to alleviate the inequalities surrounding the distribution of genetic resources while also redistributing the benefits from developing these same resources. Southern countries viewed the arena as an opportunity to acquire new funding needed for development and keep their natural resource base intact for future needs. Northern countries agreed upon the need to conserve habitats and ecosystems as the basis for protecting biological diversity. Many in the North also agreed that the cost to do so should not fall on the developing countries who could not afford to do so [1]. However, Northern countries did not view the inequalities in the distribution of benefits as an imperative treaty outcome.

Northern countries are the chief users of genetic material as this may be the basis for pharmaceutical research and development as well as the development of biotechnology. Historically, pharmaceutical companies accessed this genetic material for free; it had no value short of its conversion into a finished product. Additionally, international corporations invested heavily in research and development to create biotechnology. Biotechnology is any technological application that uses biological systems, living organisms, or derivatives thereof, to make or modify products or processes for specific use [43]. Biotechnology includes genetically modified organisms (GMOs), also known as living modified organisms (LMOs), including seeds that are more productive or drought resistant. Genetically engineered examples may also include animals. LMOs are commonly found in the food chain in the United States as well as other countries. USDA [54] statistics indicate that 92% of cotton, 94% of soybeans, and 94% of cotton grown in the United States originates from a GMO.

Like many other environmental problems, the question of conservation of natural resources may be thought of as a "tragedy of the commons" situation. In providing initial support for negotiating a treaty of conservation matters, developed countries, particularly the United States, envisioned a treaty that dealt with setting aside parks and reserves but did not contemplate including biotechnology or access to genetic resources [34]. However, the scope of the treaty grew significantly from a conceptualization of a conservation scheme for land and animals to a broader treaty concerned with biodiversity and genetic access as others saw these issues as intrinsically linked [14, 31].

Unsurprisingly, states turned to the same solution for this common pool resource situation, governmental regulation through the form of an international treaty. Given the relationship between prominent actors within international environmental diplomacy, conservation made its way into the WCED report on sustainability, namely, *Our Common Future*. UNEP's Governing Council, in decision 14/26, requested that the Executive Director assemble a group of experts to advise him on the "desirability and possible form of an umbrella convention to rationalize current activities in this field and to address other areas which might fall under such a convention" [45: 79].

The Executive Director, acting in accordance with this directive, created the Ad Hoc Group of Experts to the Executive Director of UNEP. This group conducted the initial assessment of the status of conservation efforts and the treaties that provided some assistance in protecting the environment. From the onset, UNEP and its consultants recognized that several important treaties covered some areas of conservation, including endangered species by the Convention on International Trade in Endangered Species of Wild Fauna and Flora, ecosystem protection under the Convention on the Conservation of Antarctic Marine Living Resources and migratory species of birds through the Convention on the Conservation of Migratory Species of Wild

Animals among others. However, the piecemeal approach to conserving biodiversity left large swaths of environmental resources exposed to the whims of countries, corporations, and individuals.

In its Report on the Work of its First Session, held from November 16–18, 1988, the Ad Hoc Group of Experts to the Executive Director of UNEP (Expert Group) addressed multiple concerns about conservation efforts. Over the course of the next two years, the Expert Group convened three times [46]. This group suggested that UNEP identify elements for incorporation into a draft text for a new treaty for consideration by all UN member states. States reasoned that attempting to create a new treaty by amending any or all the conventions created a logistical nightmare as each treaty had been accepted by differing groups of states. As no treaty had been universally ratified, recommending the utilization of any of the existing treaties would come at the risk of offending at least one country.

The creation of a new, standalone treaty raised significant questions that would need to be answered early in the negotiating process. First, states needed to decide whether the biological diversity treaty would utilize the framework-protocol mechanisms or contain substantive text. Second, states would need to determine the relationship of the new treaty with the older treaties that previously entered into force. While the Expert Group did not want to disturb existing structures that managed specific aspects of biodiversity, states highlighted the need for efficacy and coordination between the various convention secretariats. Further, the Expert Group noted that other efforts in addition to international treaties would likely be necessary as an over-arching treaty would not cover the entire range of conservation activities.

The experts convened in various settings to gather information about biodiversity as neither scientists nor states possessed complete knowledge about the current rates of decline of biodiversity at the individual species or ecosystem levels. This data gap exacerbated attempts to apply the environmental rationale of assessment, management, and supporting measures developed at UNCHE in Stockholm in 1972, as states face difficulties managing unknowns. Additional information requested included financial estimates as states needed to understand the number of ecosystems, their locations, and the complexity of the tasks. Thus, additional reports were commissioned on important topics such as financing mechanisms for conservation, funding sources, biotechnology, and the relationship between intellectual property rights, genetic access, and biotechnology development [46].

UNEP's Executive Director reported on the work of the Expert Group to the UNEP Governing Council in its fifteenth session in May 1989. As part of its Decision 15/34, the Governing Council confirmed the work of the Expert Group [47]. This Decision also raised awareness about IUCN's draft treaty articles so that states might consider them for inclusion. The inclusion of an NGO in this decision is itself noteworthy and undoubtedly increased the prestige of this organization. The decision also authorized the Ad Hoc Group of Legal and Technical Experts on Biological Diversity (AGLTE) to begin meeting upon completion of the work of the Expert Group to finalize a draft negotiating text for consideration as soon as possible. While a deadline for this work does not appear in the Decision, Executive Director Tolba was requested to provide an update by the PrepCom I for UNCED.

IUCN finalized its suggested text in June 1989. The text focused more on the intrinsic value of wildlife and upon conservation of its habitat [13]. IUCN demonstrated a shift in thought away from a purely conservation-focused document by attempting to incorporate Southern development needs. The document made the radical suggestion that Northern countries pay to access the South's genetic resources in Article 28 [13]. Southern countries readily endorsed this conceptualization, while not surprisingly, Northern countries resisted. Thus, the North–South gap took center stage while conceptualizing the treaty's content. Value conflicts included whether to situate the treaty text within the common heritage of humankind framework or state sovereignty principles. Northern states viewed the treaty as an economic issue focused upon genetic resources. Swanson [40] framed the need for a CBD as an economic problem stemming from countries undervaluing their natural resources. He argued for expanding the patent system to cover genetic materials in both national and international law. This system would increase the value of natural resources in its unaltered state and facilitate agreements based upon the existing international patent system.

Alternatively, Southern states viewed the problem as another attempt to stop them from developing and, therefore, a violation of Southern state sovereignty. Kothari [14] pointed out that portions of the developing South remained concerned about the CBD serving as a back door implementation of the climate change regime being developed on a parallel track. He was concerned that Northern states would require Southern states to protect their forests as a carbon sink. This potential action could also protect some of the South's biodiversity, as it may be found in tropical forests such as the Amazon. However, the requirements to preserve this tropical forest and other forests potentially eliminated one pathway for future economic development for Southern countries.

The Expert Group met twice more in 1990 before finalizing its work. During the second session, held in Geneva, Switzerland, from February 19–23, 1990, experts realized that the common heritage of humankind might not apply to this situation as the term historically referred to common pool resources beyond national jurisdictions. This is not the case for the contemplated treaty as the common pool resources lie within the state's jurisdictions [49]. Some states also pointed out the necessity of linking conservation with resources to complete the task. Additionally, states settled the question of the form of the treaty when they decided to create a new international treaty of conservation, including both in situ and ex situ mechanisms.

The Expert Group returned to Geneva from July 9–13, 1990, to complete their work. States reviewed previously commissioned studies estimating the cost of conservation. Estimates presented at the meeting varied but typically ranged from $1 billion to $50 billion per year in additional funding [48]. These figures represented a dramatic increase in development assistance, triggering robust conversations about future funding sources outside of formal development assistance. States also raised the possibility of using the newly created GEF as a potential funding source.

The completion of the Expert Group allowed the Ad Hoc Working Group of Legal and Technical Experts on Biological Diversity (AGLTE) to begin reviewing draft treaty text covering the primary issues identified previously: conservation of natural resources, fair access to genetic resources and technology, and funding for conservation and administrative costs. AGTLE met in its first session from November 19–23, 1990, in Nairobi. The AGTLE discussed principles that should tie the document together and agreed that the common heritage of humankind should not be used to describe biodiversity.

Instead, this group also adapted the common concern of humankind as an emerging principle distinct from the common heritage of humankind during these conversations [37]. The recognition of state sovereignty differentiates these two concepts. In the common heritage of humankind, no one group owns the natural resource. In declaring the common concern of humankind as the guiding principle for biodiversity, states recognized national jurisdiction while also seeking balance in competing interests in decision-making for maintaining, using, and distributing the benefits from biodiversity. States also sought to incorporate intergenerational equity in line with sustainable development. This principle is an inherent trait of conservation. States also reviewed other draft articles of the text. As is typical for these types of negotiations, a few items were finalized during this meeting.

The AGTLE reconvened from February 23–March 6, 1991, also in Nairobi for its second session. Occurring against the backdrop of the Iraqi Gulf War, the Working Group demonstrated deep divisions, including the inability to agree on a Bureau to organize and run the session. States expressed concern about the origins of the working draft treaty used as the basis for negotiations as IUCN seconded staff to UNEP to write the draft treaty language. Official documents indicated a preference to only include draft text based upon the instructions of states [52].

During this second session, states created smaller groups in order to begin the process of negotiating the draft treaty text in time for the UNCED conference in 1992. Sub-Working Group I reviewed essential concepts for inclusion in the treaty, including conservation of biological diversity for present and future generations, unique considerations of developing countries' circumstances, funding mechanisms, and the elucidation of a viable relationship between conservation, development, and sustainability. The negotiation of international treaties generally identifies key principles then works to create meaningful text from these principles. Sub-Working Group II organized their work by substantive issues, including the scope and purpose of accessing biological diversity, principles that should govern access, and measures of access.

In essence, Sub-Working Group II focused on the inclusion of all types of genetic materials regardless of whether the material came from plants, animals, or micro-organisms, irrespective of whether the material was located in a territory under national jurisdiction or part of the global commons. States also defined the purpose of access broadly, including conservation, rational, sustainable, or economic development, and scientific research. The purpose not only included direct possession of genetic materials but also access to information.

While more controversial, states also discussed the idea of regulating access to biodiversity. On the one hand, Southern states asserted state sovereignty in protecting diverse areas under their national jurisdiction. On the other hand, restricting access to these sites meant that Southern countries would become burdened with maintaining the area with no external funding to help support these efforts. Thus, one of the critical issues became how developed countries could assist developing countries with funding in order to preserve these valuable ecosystems without jeopardizing the developing countries' ability to continue their economic development in the future.

Measures of access to biodiversity included conditions about how states would share information about biodiversity. Items discussed included a clearinghouse for information and research and a database about uses of biodiversity. Joint research initiatives were proposed to close the gap regarding scientific exploration and the ability to create new biotechnologies and conditions for sharing technologies developed from the research. These items proved highly controversial as states debated whether pharmaceutical research should be included in the technology transfer. Diplomats also pointed out that organizations such as the General Agreement on Tariffs and Trade (GATT), the precursor organization to the WTO, or the World Intellectual Property Organization could also be considered an appropriate forum for conversations given the technology transfers concerned ownership of intellectual property rights. Both of these alternative organizations could reasonably be expected to champion the rights of the patent holders over other competing interests.

States requested to rename the AGTLE as an International Negotiating Committee to avoid confusion with other groups with similar names moving forward after this meeting. The UNEP Governing Council granted this request in May 1991 [53]. Thus, the INC for a Convention on Biodiversity came into existence. These states also recognized the need to coordinate their efforts with the PrepCom for the UNCED.

The International Negotiating Committee met five additional times over the course of the following year, a surprisingly rapid turn of events for international negotiations undoubtedly driven by states' desire to complete negotiations in time to present the treaty at the Earth Summit in 1992 [21]. Roberts [29] noted that this process meant that the treaty contents weakened in terms of states' responsibilities and outcomes based upon the need to complete the work. It also encouraged states to utilize existing international infrastructure for new purposes rather than creating new infrastructure.

9.2   The CBD

The CBD opened for signature during the Earth Summit in 1992. The CBD accurately reflected the desires of the South for determining the terms of access for genetic materials that were under their national jurisdiction. Article 1 of the CBD detailed three important elements: biodiversity conservation, the sustainable use of resources, and the equitable sharing of benefits [43]. As with many other international treaties, the CBD also determined an operational support structure and funding mechanisms and provided specific guidelines for organizational matters.

The treaty does suggest that biodiversity should be preserved both in situ (in nature) and ex-situ (off site). Thus, biodiversity protection should apply to ecosystems as a protected area and through collecting materials such as germplasm that may be stored in a seed bank. Therefore, the CBD intended to include species diversity and ecosystem diversity. While at the beginning of the treaty negotiations, states focused on protecting habitat to save individual species but changed their focus to ecosystems services [26]. However, the CBD did not include specific details on how either of these two interrelated outcomes should occur as it fails to identify any individual site, species, or ecosystem for inclusion within a protection system [22]. Instead, the treaty text left this detail to member states for consideration.

The CBD also set forth considerations regarding the sustainable use of valuable bioresources. States linked the conservation component in the treaty text to benefit sharing. States that preserved their diverse ecosystems exercised their state sovereignty to demand a significant shift in terms of access. The freedom of access to genetic resources system that favored Northern counties in the past was replaced with a new system that demanded benefit-sharing as a condition of access.

The CBD is not a preservation treaty that does not allow any type of use of the area. Instead, the CBD allows the sustainable use of resources. That is, resources may be used as long as the Earth can replace the resource. The harvesting of resources, such as fish, may occur if the population has the ability to regenerate. Regeneration rates depend on the resource being consumed, as some natural resources regenerate faster than others. Müller [25] points out that limiting states' ability to use their resources for sustainable uses only is technically a restriction on state sovereignty.

The CBD established a quid pro quo situation in that some type of transfer should take place in exchange for access. Options identified include financial compensation, joint participation in the research endeavor, ownership of intellectual property rights, manufacturing facilities established in the developing countries, and patent rights to any new drugs, seeds, or other biotechnology produced. This extensive list represented a clear economic, social, and environmental victory for the South. This circumstance, however, was mitigated by inserting the phrase that these transfers would agree upon the terms and conditions by which the transfers occurred.

The CBD acknowledged the necessity of increasing development aid to achieve these goals through the creation of new and additional funding for biodiversity. This treaty created an additional regulatory burden on states who must now intercede on behalf of the natural environment and these states may not have the resources to pay for the new regulatory system. While the CBD theoretically established mechanisms for transferring resources from the North to the South, this only occured when a Northern interest wishes to access Southern biodiversity. Given the dictates of this system and the likelihood that Northern research and development labs have their own genetic stockpiles, access under the treaty may not occur immediately. Southern countries, however, will accrue expenses for establishing the regulatory scheme in advance of receiving a payment under the plan. To meet the needs for funding, a wide variety of sources may be called upon. States requested that GEF serve as the interim funding mechanism while states created a permanent mechanism as specified in Article 21. The CBD also called for the creation of a Secretariat to provide administrative support and coordinate activities with other international bodies.

On December 29, 1993, the CBD entered into force after thirty countries ratified the document. However, the United States was not among these countries. The benefit sharing provisions contained within the CBD ultimately led to the United States unabashedly refusing to sign the treaty. Blomquest [2] provided extensive research into the public statements of the Senators that would potentially be responsible for approving this treaty. The concerns expressed during the run-up to Rio involved subjecting the United States pharmaceutical companies to international scrutiny, opposition to potential pre-import approvals from other countries, and concerns about funding developing countries. Technology transfer, also, received significant scrutiny from Senators who questioned why developed countries' corporations should agree to give away biotechnologies and pharmaceutical recipes that were costly to develop.

9.3 The CBD and TRIPS

The creation of the CBD raised considerable controversy, especially surrounding patents. Thus, signing the CBD meant that states agreed to manage their patent system on a set of principles without knowing the extent to which a patent system would be allowable in the future. This highly contentious issue linked together with other international organizations and the regimes they administered, such as the GATT/WTO oversight of the Trade-Related Aspects of Intellectual Property Rights (TRIPS) (e.g., [30]). TRIPS updated the copyrights and patent system for diverse industries such as the entertainment and music industries as well as biotechnologies and medical technologies. Consequently, states negotiated rules for patents under the auspices of the CBD and TRIPS simultaneously.

States created GATT on October 30, 1947, to contribute to the post-World War II recovery. GATT focused on liberalizing economic trade by reducing barriers to moving goods between countries. Prior to the establishment of GATT, a highly complex and contentious system of tariffs had evolved that determined whose goods were allowed to enter a country [11]. Tariffs also determined prices as companies sought to recover the tariff from the consumer. Most Favored Nation (MFN) status, one of the most important of the GATT principles, represents a specific level within international trade that may include a low tariff rate, a high level of imported goods allowed into a country, or both. Upon joining GATT, states agreed to confer MFN status on other GATT member states. GATT member states reciprocated by granting MFN status to the new member state. Regional trading blocs such as the EU receive an exemption from MFN rules. Additionally, least developed countries (LDCs) may be completely exempt from both import duties and import quotas.

States gained interest in developing a treaty about intellectual property rights and violations of these rights, including counterfeiting and piracy, by negotiating the TRIPS that entered into force in 1995. Joining TRIPS is mandatory for countries participating in the WTO. While states may not agree with the content of TRIPS, forgoing this treaty also meant forgoing MFN status, creating a significant economic disincentive for opting out of this treaty. As an example of the impact of TRIPS, around 50 countries did not allow pharmaceuticals to be patented domestically, and this situation would need to change [44]. Subhan [39] pointed out that the uneven enforcement of copyrights and patents negatively impacted innovation as the copyright violator typically undercut the original price and possibly made more money in the process. TRIPS was not the first international agreement on intellectual property. The Paris Convention on the Protection of Industrial Property of 1883 also covered some of the same ground. TRIPS did, however, add new requirements by requiring states to guarantee a minimum life of twenty years for a patent. It also extended the patent to cover the manufacturing process, thereby closing a loophole for legal reproductions of patented innovations.

TRIPS' Article 27.1 stated that "patents shall be available for any inventions, whether products or processes, in all fields of technology, provided that they are new, involve an inventive step and are capable of industrial application" [56]. Thus, TRIPS required countries to allow patents to genetic materials that may not have received protection previously. States that did not wish to create a formal patent system had the option of providing a *sui generis* (unique) coverage that provided equivalent protection per Article 27.3(b) [56]. Article 27.2 provided an exemption for the protection of the environment [56]. However, the exemption may not be invoked merely because of the existence of domestic law. In other words, states needed a valid reason to avoid issuing a patent, other than the fact that the state did not wish to do so. In this sense, TRIPS reflected the concerns of developed countries more closely, especially regarding patenting of medical technologies and pharmaceuticals.

TRIPS attempted to balance the inclusion of an expanded patent system by creating compulsory licensing and parallel importing to address public health concerns of developing countries. These countries convincingly argued that the differences in development meant that residents in developing countries were less likely to be able to access life-saving medical resources, including technologies and medicines. Creating unattainable conditions of access resulted in preventable deaths within these countries. TRIPS' Article 31 contains the conditions for establishing compulsory licensing. Compulsory licensing occurs when a state allows someone other than the patent holder to manufacture the product without the patent holder's consent. Compulsory licensing effectively allowed a corporation other than the patent holder to produce a generic version of the same drug at a significantly lower cost. The generic producer did not include the costs of research and development in its pricing. The state issuing the compulsory license determines the royalty fee the patent holder will receive.

Compulsory licensing may take place in emergency situations such as a pandemic or non-emergency situation as determined by the state. In non-emergency cases, the company seeking to produce the medicine should make a reasonable effort to purchase the patent license. TRIPS included restrictions for selling this generic brand in competition to the original patent holder. The product's sale should occur in the state that granted the compulsory license only [56]. Otherwise, TRIPS would have created a pathway that destroys the reward for innovation in contradiction to its mandate to protect intellectual property rights.

Parallel importing allowed countries to take advantage of price differentials in various countries to secure the lowest cost for the product. Parallel importing occurs when a company with the right to manufacture and sell a product does so without the permission of the patent holder. During the COVID-19 pandemic, parallel importing would allow a country like Egypt to purchase monoclonal antibodies from the United Kingdom at a lower price rather than to buy the domestically produced monoclonal antibodies from the Egyptian patent holder at a higher price.

Both provisions deeply concerned the pharmaceutical industry and the United States government. In 1997, the United States retaliated against South Africa, who implemented the provision in domestic law [7]. The United States began the process to levy unilateral trade sanctions in direct response to the South African law. This unilateral action, combined with the United States' decision to refrain from signing the CBD, signaled the importance of ensuring international recognition of a strongly patent system to this country. Southern countries felt equally strong about the necessity of implementing equitable access and sharing benefits under the CBD. Countries such as Bolivia, Colombia, Ecuador, Peru, and Venezuela required patent applications in these countries to recognize the contributions of indigenous communities and traditional knowledge holders [30].

States could therefore choose which treaty to emphasize in their domestic policies as TRIPS and the CBD create fundamentally different principles. Rosendal [30] pointed out that TRIPS created exclusive rights to genetic resources through the use of a patent system while the CBD sought to ensure equitable sharing and benefits, thereby ending the exclusive nature of the same set of patents. Unsurprisingly, states chose to emphasize the treaty closest to their own position as Northern countries such as the United States supported TRIPS over the CBD. The United States, as of January 2022, has not signed the CBD but participates in TRIPS. Conversely, Southern states supported the CBD over TRIPS (e.g., [16, 17]).

The WTO is easily the more influential of the two regimes as its compliance and enforcement mechanism is more robust and meaningful [16]. In contrast, the CBD does not possess any significant compliance and enforcement capabilities. Kothari and Anuradha [15] recommended taking advantage of the WTO's enforcement mechanism by adding the CBD's focus on equitable access and benefit sharing to the WTO TRIPS regime. The CBD Secretariat has applied for membership on the TRIPS Council at the WTO, but its application remains pending as of January 2022. The pending application of the CBD Secretariat to the TRIPS Council notwithstanding, the two international organizations have attempted to ease the tension between the two regimes; however, tensions and difficulties continue to exist.

9.4 The Cartagena and Nagoya Protocols

After opening the CBD for signature, states began the process of forming the regime. Unlike the UNFCCC that created a COP immediately, the CBD established an Intergovernmental Committee on the Convention on Biological Diversity that met before COP 1. This short-lived group operated during the period after the CBD opened for signature and before the first COP in November 1994. The group met twice, once in

October 1993 and again in June 1994. The first meeting was unremarkable, and the second meeting did not produce any significant work.

Therefore, the agenda for COP 1 contained infrastructure issues such as determining which institution should be the Secretariat, the location of the Secretariat, and designating a permanent financial institution. States also negotiated the Medium Term Work Program, defining an agenda for the Subsidiary Body on Scientific, Technical, and Technological Advice, and established an Ad Hoc Group of Experts on Biosafety to investigate the need for an additional protocol dealing with biosafety. States agreed on designating UNEP as the conference secretariat but postponed deciding upon a permanent location as multiple states expressed an interest in hosting this institution. States selected Montreal, Canada, as the home for the CBD Secretariat during COP 2.

Southern states continued their relationship with GEF on an interim basis until designating GEF as the permanent financial institution in COP 3. Developing countries distrusted GEF given its close association with the World Bank [8, 36]. Thus, the permanent designation of GEF as the financial trustee for this work became dependent on GEF reform that expanded its membership. Southern states also sought a role in determining what projects would be funded as many of the Southern states did not believe the World Bank acted in developing countries' best interests [36]. Southern countries also demanded and received guarantees that developing countries would be able to participate in the management and oversight of this group. In 1994 GEF distanced itself from the World Bank. Concurrent with the shift in structure, states needed to replenish funding. However, only half of the amount pledged to support the Earth Summit outcomes arrived. Sánchez [34] believed that donor states preferred to invest in rebuilding Eastern Europe that was in the process of transitioning from communism to democracy instead of meeting the commitments made to developing countries.

Over time, complex financing mechanisms for biodiversity emerged that resulted in the creation of multiple trust funds. The General Trust Fund for the Convention on Biological Diversity utilizes the UN scale of assessments that is based upon a country's ability to make contributions. Two additional trust funds exist to help defray costs, including supporting member states to attend meetings. Additionally, states may (and do) make additional voluntary contributions to the trust funds. Today, GEF remains actively engaged in providing funding for environmental projects, including biodiversity.

During COP 2, states determined that a need existed to create a protocol on biosafety and organized an Open-Ended Ad Hoc Working Group on Biosafety to negotiate the text. This treaty could potentially cover items such as seeds that could be planted or, in the formal words of the treaty, LMOs intended for release into the environment or for other agricultural commodities that utilized biotechnology. Like other treaties that use the framework-protocol convention, negotiation of specific protocols may significantly alter the terms and conditions of the framework for conversation in both positive and negative manners. Schnier [35] provided a board overview of the five negotiating blocks leading up to the Cartagena Protocol. He noted that the negotiating groups varied from the traditional North–South gap as states realized that negotiating a protocol on biosafety potentially impacted their domestic industries.

These five groups occurred due to fractures that primarily occurred in the Northern developed countries. With its strong desire to avoid LMOs due to the precautionary principle, the EU diverged sharply with the United States that embraced these same resources [6, 35]. Thus, the United States allied with Latin American and Caribbean countries that also focused on promoting agricultural exports or otherwise perceived a promising future in developing biotechnologies. The so-called Miami Group concentrated on protecting these industries from what was, in their view, unnecessary and unsupported trade restrictions on LMOs [35, 53]. The Like-Minded Group, consisting of the majority of the G-77 countries and China, formed the largest group of countries in the negotiations and argued consistently for regulations that fully implemented the precautionary principle by limiting LMOs, including notifying countries in advance of a shipment and providing for a strong liability system [35]. Additionally, the Central and Eastern European Group consistently aligned with the Like-Minded Group. The last group, the Compromise group, consisted of developed countries that attempted to broker a deal between these radically different positions [35].

The Open-Ended Ad Hoc Group of Experts convened six times, beginning in Aarhus, Denmark, in July 1996. By 1998, signs emerged that this group might not complete its work on schedule. COP 4, in May 1998, foreshadowed the difficulties states experienced in reaching consensus on the content of the Cartagena Protocol as states altered the negotiating deadline to give themselves additional time to finalize the treaty text. States intended to finalize the Protocol at the Sixth Meeting of the Open-Ended Working Group on Biosafety and Extraordinary Session of the COP in Cartagena, Columbia, in February 1999 that met for this purpose. However, the meeting proved more difficult and contentious than anticipated. States strongly disagreed about which products should be included in the Protocol. Additionally, they also could not agree upon what products should be regulated by the Protocol, the usage of the Advanced Informed Agreement procedures, and the relationship between parties to the Cartagena Protocol and non-parties [10].

In order to avoid declaring the meeting a failure, diplomats suspended the session and completed the negotiations a year later in Montreal, Canada, in January 2000. The Cartagena Protocol added to the CBD by regulating LMOs that could negatively impact the environment. While the Protocol does not ban the use of LMOs, the protocol does impose additional regulations upon entities seeking to transport these goods across national boundaries. Thus, LMOs that jeopardize biodiversity must seek Advanced Informed Agreement before shipping. However, LMO products intended for direct consumption (food, feed, or medicine) are included in the Cartagena Protocol but exempted from the Advanced Informed Agreement notification regulations.

States receiving these products utilize the Advanced Informed Agreement procedure to review and approve the first intentional introduction of the LMO into the environment. After receiving the information, the importing party must acknowledge receipt within 90 days. The importing state must then decide within 270 days, whether to permit the LMO including a reason for the rejection or extending the review period. The Cartagena Protocol allows states to reject products based upon the presence of GMOs [10, 35]. The importing party must justify its decision, but the decision may take into account scientific assessments, the precautionary principle, and relevant socioeconomic considerations. The Cartagena Protocol also contains provisions for a Biosafety Clearing House, an information exchange platform for interested parties seeking information about LMOs. Under the Advanced Informed Agreement Procedure, importing parties must create a record of their decisions and make the document publicly available through the Biosafety Clearing House.

In a noteworthy departure from many other international environmental treaties, the Cartagena Protocol is legally binding. States that ratify the Protocol must also implement the Protocol. Similarly, states that wish to participate in the Cartagena Protocol must also ratify and implement the CBD. Consequently, the United States will not be allowed to become a party to the Cartagena Protocol until its ratification status changes with regards to the CBD. With these compromises in place, states adopted the Cartagena Protocol on January 29, 2000. After receiving the fiftieth ratification, the Cartagena Protocol entered into force on September 11, 2003.

The Nagoya Protocol intended to operationalize the benefit-sharing provisions contained within the CBD. Due to the difficulty of the negotiations and the tight deadline for completing a treaty in time for the Earth Summit, the CBD did not contain details to operationalize all the principles. Thus, the Nagoya Protocol, like the Cartagena Protocol, added a layer of specificity to the pre-existing treaty. Morgera et al. [24] framed the need for the Nagoya Protocol around the theoretical concept of asymmetries, the unequal distribution of a resource. These asymmetries may be represented by the North–South gap. They also include asymmetries in the existing distribution of biodiversity between countries.

COP 5 established an Ad Hoc Open-Ended Working Group on Access and Benefit Sharing (Working Group on ABS) in May 2000. Initially, this group focused on providing advice for interested parties to participate in the access and benefit sharing provisions of the CBD. The group met twice and produced the Bonn Guidelines on Access to Genetic Resources and Fair and Equitable Sharing of the Benefits Arising out of their Utilization. As these guidelines are non-binding, states agreed to negotiate a legally binding protocol at COP 7 in 2004. Thus, the Working Group on ABS began work on a mechanism that would become the Nagoya Protocol with a mandate to report on their progress at COP 8 in March 2006 [52].

Establishing further direction for the Working Group on ABS occurred at COP 8 in Curitiba, Brazil, as states debated which document to use as the basis for the negotiating text for the Protocol. The COP also confirmed that mutually agreed terms should be part of the PIC process in cases where genetic resources may be utilized, or traditional knowledge may be involved. Additionally, states determined to complete their work and produced a draft text by COP 10.

The Nagoya Protocol assisted in determining the terms of access to genetic resources. Scientists, manufacturers, or others interested in utilizing genetic resources preserved by indigenous peoples should negotiate mutually agreed terms, i.e., a private contract, agreeable to both parties. These terms may include monetary and non-monetary forms of compensation as specified in the Annex to the treaty. Potential non-monetary forms of compensation that could be included in a package include research and development results, commercial ownership, or other benefits. However, the protocol makes no recommendation regarding the specific packages offered.

States are required to create a PIC process to comply with the Nagoya Protocol. States should not utilize this process as an arbitrary mechanism to prevent interested parties from accessing genetic resources. As part of the PIC process, parties wishing to access the genetic resource should provide documentation of the mutually agreed terms, as part of the documentation. Additionally, parties interested in accessing genetic resources may be asked to pay a fee for the registration of the PIC permit.

Further terms and conditions for the PIC process may apply. For example, parties granting access to genetic resources may provide model language to include in the terms of the contract. As with any other regulatory scheme, states may impose penalties or jail time for non-compliance with the PIC process. In order to create transparency around the PIC process, the Nagoya Protocol created an Access and Benefit Sharing Clearing House on the CBD's website. Parties to the Nagoya Protocol are required to post information about their specific PIC process, national contacts, and permits (or equivalent) issued in the Clearing House. Anyone wishing to access the Access and Benefit Sharing Clearing House online may do so. Prathapan and Rajan [28] noted that the outcome of the Nagoya Protocol with its emphasis on requiring national permits with an embedded requirement to negotiate private law contracts for access and benefit-sharing commercialized genetic resources instead of retaining focus on an environmental need for conservation. Nevertheless, states finalized the Nagoya Protocol on October 29, 2010. The Nagoya Protocol entered into force on October 12, 2014, 90 days after the fiftieth ratification. As of January 2022, 132 countries have ratified the Nagoya Protocol.

9.5 Analysis

Compared to its sister treaty, the UNFCCC, the CBD suffered from a lack of attention [41]. While the more successful of the two treaties in institution-building and environmental management, the CBD remains less popular than the climate change regime. Ironically, McGraw [21] points out that the high point in popularity for this issue area was undoubtedly the moment in 1992 when the United States announced its refusal to join other countries in ratification.

That is not to say that scholarly analysis of the CBD does not exist; it does. Academic scholarship initially focused upon the dual nature of the CBD as both an environmental treaty and an economic treaty. Downes [5] referred to the CBD as a trade agreement for genetic resources that, while significant as a first attempt, nevertheless required refinement. This treaty also contained conservation methods necessary to preserve the future productivity of this highly valuable commodity. LePrestre [18] agreed that the CBD serves to combine elements of a trade agreement and a conservation treaty but instead viewed the CBD as establishing a process for handling issues linked to biodiversity.

Another strand of scholarship reviewed the linkages between environment and economics. The research focused on reconciling the differences between TRIPS and the CBD. While Rosendal [30] reviewed the interlinkages between TRIPS and the CBD without making policy suggestions, Sahai [32] recommended that TRIPS be amended to promote the CBD. Other scholars point out that removing biodiversity from TRIPS also reconciles this issue [15, 33]. Morgera et al. [24] suggested utilizing the WTO's stringent dispute resolution process to potentially strengthen the CBD as it does not have an equivalent provision.

Separate from analyzing economic issues, Tinker [41] praised the treaty for its incorporation of cultural diversity as an important component of biodiversity. While Tinker [42] noted the treaty could be considered significant due to its broadening of the actors within international environmental politics as this treaty considers the role of indigenous people and women in protecting biodiversity. These groups stand to benefit directly from the implementation of the CBD regulatory system, assuming that the mechanisms for the mutually agreed terms are implemented by states and utilized in practice.

However, the international state system has proceeded slowly toward this mechanism due to the highly controversial nature of changing the fundamental principles for accessing biodiversity. Further, the complexity of the topic requires coordination across multiple international institutions, at least one of which is not known for its environmental sensitivity. Consequently, hopes for rewarding the developing countries and their indigenous peoples may remain far off in the future.

References

1. Barton JH (1992) Biodiversity at Rio. Bioscience 42(10):773–776
2. Blomquist RF (2002) Ratification resisted: understanding America's response to the convention on biological diversity 1989–2002. Golden Gate UL Rev 32(4):493–586
3. Butler R (2020) The Amazon rainforest: the world's largest rainforest. Mongabay https://rainforests.mongabay.com/amazon/. Accessed on 27 Nov 2021
4. Contos J, Tripcevich N (2014) Correct placement of the most distant source of the Amazon River in the Mantaro River drainage. Area 46(1):27–39
5. Downes DR (1994) The convention on biological diversity: seeds of green trade? Tulane Environ Law J 8(1):163–180
6. Falkner R (2000) Regulating biotech trade: the Cartagena protocol on biosafety. Int Aff 76(2):299–313
7. Fisher W, Rigamonti C (2005). The South Africa AIDS controversy: a case study in patent law and policy. https://cyber.harvard.edu/people/tfisher/South%20Africa.pdf Accessed on 23 Jan 2022
8. French HF (1994) Strengthening international environmental governance. J Environ Dev 3(1):59–69
9. Gatti LV et al (2021) Amazonia as a carbon source linked to deforestation and climate change. Nature 595:388–393
10. Hagen PE, Weiner J (2000) The Cartagena protocol on biosafety: new rules for international trade in living modified organisms. Georget Int Environ Law Rev 12(3):697–716

11. Irwin DA (1995) The GATT in historical perspective. Am Econ Rev 85(2):323–328
12. IUCN (1980) World conservation strategy living resource conservation for sustainable development. IUCN, Gland
13. IUCN (1989) Draft articles prepared by IUCN for inclusion in a proposed convention on the conservation of a biological diversity and for the establishment of a fund for that purpose with explanatory notes. https://portals.iucn.org/library/sites/library/files/documents/Rep-1989-094_0.pdf. Accessed 10 Jan 2022
14. Kothari A (1992) Politics of biodiversity convention. Econ Polit Wkly 27(15/16):749–755
15. Kothari A, Anuradha RV (1997) Biodiversity, intellectual property rights, and GATT agreement: how to address the conflicts? Econ Polit Weekly 32(43):2814–2820
16. Kothari A, Anuradha RV (1999) Biodiversity and intellectual property rights: can the two co-exist? J Int Wildl Law Policy 2(2):204–238
17. Laxman L, Haseeb Ansari A (2012) The interface between TRIPS and CBD: efforts towards harmonisation. J Int Trade Law Policy 11(2):108–132
18. LePrestre PG (2017) Introduction the emergence of biodiversity. In: LePrestre PG (ed) Governing global biodiversity: the evolution and implementation of the convention on biological diversity. Routledge, London, pp 1–6
19. Lovejoy T (1984) Aid debtor nation's ecology. NY Times A 31
20. McCormick J (1986) The origins of the world conservation strategy. Environ Rev 10(3):177–187
21. McGraw DM (2017) The story of the biodiversity convention: from negotiation to implementation. In: LePrestre PG (ed) Governing global biodiversity: the evolution and implementation of the convention on biological diversity. Routledge, London, pp 7–38
22. McNeely JA, Rojas M, Martinet C (1995) The convention on biological diversity: promise and frustration. J Environ Dev 4(2):33–53
23. Moran EF (1993) Deforestation and land use in the Brazilian Amazon. Hum Ecol 21(1):1–21
24. Morgera et al (2015) Unraveling the Nagoya protocol: a commentary on the Nagoya protocol on access and benefit-sharing to the convention on biological diversity. Brill, Leiden
25. Müller FG (2000) Does the convention on biodiversity safeguard biological diversity? Environ Value 9(1):55–80
26. Perrings C (2010) The economics of biodiversity: the evolving agenda. Environ Dev Econ 15(6):721–746
27. Philips T (2019) 'War for survival': Brazil's Amazon tribes despair as land raids surge under Bolsonaro. The Guardian, https://www.theguardian.com/world/2019/oct/02/war-for-survival-brazils-amazon-tribes-despair-as-land-raids-surge-under-bolsonaro. Accessed 27 Nov 2021
28. Prathapan KD, Rajan PD (2011) Biodiversity access and benefit-sharing: weaving a rope of sand. Curr Sci India 100(3):290–293
29. Roberts P (1992) International funding for the conservation of biological diversity: convention on biological diversity. Boston Univ Int Law J 10(2):303–349
30. Rosendal GK (2006) The convention on biological diversity: tensions with the WTO TRIPS agreement over access to genetic resources and the sharing of benefits. In: Oberthür S, Gehring T (eds) International interaction in global environmental governance: synergy and conflict among international and EU policies. MIT Press, Cambridge, pp 79–102
31. Ryder OA (1995) Zoological parks and the conservation of biological diversity: linking ex situ and in situ conservation efforts. J Environ Dev 4(2):105–120
32. Sahai S (2001) TRIPS review: basic rights must be restored. Econ Polit Wkly 36(31):2918–2919
33. Sahai S (2004) TRIPS and biodiversity: a gender perspective. Gend Dev 12(2):58–65
34. Sánchez V (1994) The convention on biological diversity: negotiation and contents. In: Sánchez V, Juma C (eds) Biodiplomacy genetic resources and international relations. ACTS Press, Nairobi, pp 7–18
35. Schnier DJ (2001) Genetically modified organisms and the Cartagena protocol. Fordham Environ Law J 12(2):377–415
36. Sharma SD (1996) Building effective international environmental regimes: the case of the global environment facility. J Environ Dev 5(1):73–86

37. Shelton D (2009) Common concern of humanity. Environ Policy Law 39(2):83–86
38. Silva Junior CHL et al (2021) The Brazilian Amazon deforestation rate in 2020 is the greatest of the decade. Nat Ecol Evol 5(2):144–145
39. Subhan J (2006) Scrutinized: the TRIPS agreement and public health. McGill JM 9(2):152–159
40. Swanson TM (1992) Economics of a biodiversity convention. Ambio 21(3):250–257
41. Tinker CJ (1994) Introduction to biological diversity: law, institutions, and science. Buff J Int Law 1(1):1–25
42. Tinker C (1995) New breed of treaty: the United Nations convention on biological diversity. Pace Environ Law Rev 13(1):191–218
43. UN (1992) Convention on biological diversity (text with annexes). United Nations, New York
44. United Nations Conference on Trade and Development (1996) The TRIPS agreement and developing countries. UNCTAD, Geneva
45. UNEP (1987) Report of the governing council on the work of its fourteenth session. A/42/25
46. UNEP (1989a) Report on the ad hoc working group on the work of its first session. UNEP/Bio.Div.1/3
47. UNEP (1989b) Report of the governing council on the work of its fifteenth session. A/144/25
48. UNEP (1990a) Biological diversity: global conservation needs and costs. UNEP/Bio.Div.3/3
49. UNEP (1990b) Report on the ad hoc working group on the work of its second session in preparation for a legal instrument on biological diversity of the planet. UNEP/Bio.Div.2/3
50. UNEP (1991a) Ad hoc working group of legal and technical experts on biological diversity on the work of its second session. UNEP/Bio.Div/WG.2/2/5
51. UNEP (1991b) Proceedings of the governing council at its sixteenth session. UNEP/GC.16/27, 30 June 1991
52. UNEP (2004) Decision adopted by the conference of the parties to the convention biological diversity at its seventh meeting. UNEP/CBD/COP/DEC/VII/19
53. US brings support to biosafety protocol negotiations on GMOs (1999) Inside Washington's FDA Week 5(6):20–20
54. USDA (2020). Recent trends in GE adoption. https://www.ers.usda.gov/data-products/adoption-of-genetically-engineered-crops-in-the-us/recent-trends-in-ge-adoption.aspx. Accessed 23 Jan 2022
55. WWF (2022) Inside the Amazon. https://wwf.panda.org/discover/knowledge_hub/where_we_work/amazon/about_the_amazon/. Accessed 3 Jan 2022
56. WTO (1994) TRIPS: agreement on trade-related aspects of intellectual property rights. 15 Apr. 1994, Marrakesh Agreement Establishing the World Trade Organization, Annex 1C, 1869 U.N.T.S. 299, 33 I.L.M. 1197

Chapter 10
Limiting Exposure to Toxic Chemicals

Abstract Environmental health risks stem from the production and transportation of hazardous chemicals in addition to the toxic trade in hazardous waste. States created an international regulatory regime to control the movement of these chemicals by imposing prior informed consent before importing a hazardous chemical to another country by negotiating the Rotterdam Convention on the Prior Informed Consent Procedure for Certain Hazardous Chemicals and Pesticides in International Trade. States also created a third treaty to limit the production and use of pesticides with known human health and environmental risks through the Stockholm Convention on Persistent Organic Pollutants. While these treaties provide mandatory requirements on the movement of hazardous chemicals, they do not specify the behaviors of manufacturers in manufacturing and selling additional chemicals. Consequently, voluntary guidelines and codes of conduct also seek to limit manufacturers behaviors.

Keywords Hazardous waste regime · Rotterdam Convention on the Prior Informed Consent Procedure for Certain Hazardous Chemicals and Pesticides in International Trade · Stockholm Convention on Persistent Organic Pollutants · Persistent organic pollutants · Toxic trade · Prior informed consent

Concerns about hazardous chemicals entering the environment significantly predate their regulation through international treaties. Countries around the world recognized that undesirable chemicals entered the atmosphere not only from accidental releases and as byproducts of the manufacturing processes, they were also deliberately manufactured and transported to individuals that did not have the information needed, or the governmental infrastructures in place, to protect them from these hazards. The seriousness of the issue was so high that the reoccurring pattern was noted in the UNEP report to the UN GA when developing countries called out developed countries for using the South as a dumping ground for the North's unwanted products [44].

This pattern of product dumping occurred in the past and continues to occur despite states' efforts to end the practice. How to regulate chemical products from the United States or European markets that had been removed from the domestic market, then sold to unsuspecting citizens in developing countries, became an international question. One relevant example stems from DDT, an infamous pesticide closely associated with the environmental movement. DDT, a synthetic insecticide, began commercial development immediately after World War II. Initially, the chemical was utilized to protect Allied troops from diseases carried by insects, especially in tropical climates. Thus, DDT helped lower infection rates from diseases such as malaria or typhoid fever.

Rachel Carson wrote about the health impacts of this pesticide in her monograph, *Silent Spring* [6]. DDT not only killed insects; it also significantly altered the balance of nature by causing widespread environmental damage. The indiscriminate killer not only targeted insects it also impacted the reproductive cycles of small mammals. Further scientific research that occurred after the shocking testimony of Carson's *Silent Spring* confirmed that DDT could induce sickness in humans and is a suspected carcinogen.

Even though the United States banned the use of DDT in 1972, chemical manufacturers continue to produce and export it as the chemical still serves a useful purpose. Nor is DDT the only pesticide banned in the United States that is nevertheless manufactured in the country and exported to developing countries [8]. DDT manufacturing occurs, in part, because DDT effectively kills mosquitos that carry malaria. Malaria remains a consistent health threat for people living in or traveling to, tropical ecosystems. WHO [65] estimated that 241 million cases of malaria occur globally with an estimated 627,000 deaths from the disease in 2020. Van den Berg [58] estimated that fourteen countries utilized DDT from 2000 to 2007 to limit diseases and that several other countries contemplated allowing this chemical's use. In an updated study in 2017, van den Berg [59] noted that at least thirteen countries actively sprayed DDT from 2001 to 2014.

Consequently, people that live in malaria-prone areas may suffer from high local concentrations of DDT. In producing a recent literature review regarding the presence of environmental hotspots in Africa, Fuhrimann et al. [20] identified five hotspots on the continent. Of the 469 study sites included in the research, 86% of the studies included DDT. Notably, these studies occurred from 2006 to 2021, after the negotiation of a new international treaty to regulate the export and use of this chemical.

Additionally, the bioaccumulation patterns from this chemical also jeopardize human health in other places as DDT persists in living organisms and the environment. Kuhnlein and Chan [30] noted the presence of heavy metals such as arsenic, lead, and mercury among Indigenous Peoples of the North. Similarly, they also studied organochlorines including, but not limited to aldrin, DDT, and dieldrin, three insecticides that are known for their toxic nature and their long-range movement. Scientists utilize the surprising presence of these chemicals in Indigenous Peoples as an indicator of the movement of these hazardous chemicals.

Given the hazardous nature of DDT and other chemicals that share the same dangerous characteristics and exposure pathways, states attempted to restrict or eliminate certain hazardous chemicals in 2001 as part of the Stockholm POPs. This treaty, along with the Basel Convention discussed in Chap. 6 and the Rotterdam Convention, collectively regulate the production, import, export, use, and disposal of certain chemicals that cross national boundaries.

This chapter returns to the idea of minimizing environmental health risks by reducing exposure to toxic chemicals. Section one begins by evaluating the need to limit exposure to toxic chemicals. Section two reviews the Rotterdam Convention while section three discusses the Stockholm POPs. States elected to take advantage of the synergies between the three conventions that regulate the production, transportation, and usage of hazardous materials. Thus, section four details the merging of these three administrations.

10.1 Limiting Exposure to Toxic Chemicals

Global production of toxic chemicals grew in the 1960s and 1970s. As their hazardous properties led developing countries to ban their use in domestic markets, manufacturers increasingly looked overseas for sales. Their clients, frequently farmers in developing countries, became ensnared in a vicious cycle that Weir and Schapiro [63] referred to as a circle of poison. In order to produce an agricultural crop for export to the developed countries, farmers turned to pesticides to improve crop yield, thus exposing themselves to dangerous chemicals [37].

Swaminathan [41] estimated that pesticide poisoning occurred in developing countries at a rate that is thirteen times higher than in the United States. Human health impacts from pesticides include an increased risk of cancer, endocrine disruption, and reproductive mutations [2, 19, 34, 36, 39, 43]. Exposure pathways vary but may include inappropriate handling of a pesticide during transport and/or application, storage of a pesticide in a residential area, and a lack of personal protective equipment [3, 9, 32]. In addition to the human health hazards, pesticide use also causes water pollution from point and non-point source runoff, soil degradation, and the killing of local species of plants and animals [5, 33, 60].

States seeking an increase in their exports in order to pay back rising debt possessed neither the political will nor the chemical knowledge to regulate this toxic trade. Further, this pattern emerged in virtually every developing country. The ubiquitous nature of pesticides in agriculture meant that a global market developed with various regulations around pesticide use [23]. The exposure to pesticides on food returned to the developed countries as they imported produce sprayed with pesticides that could not legally be used domestically. Therefore, this globalized trade increased the transboundary risk of human exposure [42].

In addition to realizing that individuals located far away from the place of use could suffer adverse effects from these chemicals, a series of local catastrophes also raised concerns. While the United States began regulating pesticide use with the passage of the Federal Insecticide, Fungicide, and Rodenticide Act in 1910, two additional amendments sought to improve its positive impact on human health and the environment. First, the Federal Environmental Pesticide Control Act of 1972 created a program to control the sale, distribution, and use of these dangerous chemicals within the United States. Restricted chemicals could only be applied by a certified applicator. Second, the Pesticide Registration Improvement Act of 2003 tightened standards relating to the registration of pesticides intended for use within the United States to increase the accuracy of the EPA's risk assessments. However, nothing in this act prohibits United States-based manufacturers from exporting chemicals to other countries that have not been approved or registered for use domestically. This regulatory laxity also occurred in other developed countries as Uram [57] argued that the Federal Insecticide, Fungicide, and Rodenticide Act represented the strongest set of export controls in the world during this time frame despite its weaknesses.

Consequently, states engaged in activities to manage and reduce the environmental impacts of pesticides by moving toward a regulatory system. One important action that states could take to improve the regulation of chemicals across the globe is to share information about the health hazards associated with their use [23]. Further, sharing information about chemicals and allowing states to make their own determinations is less likely to raise concerns about violating state sovereignty than an outright ban on chemicals [27]. Ross [37] argued that the absence of information about human health and environmental impacts infringed upon state sovereignty by denying them the information required to make a rational decision.

Additionally, the lack of stringent export controls on pesticides allowed manufacturers to engage in unscrupulous business practices. Exporters could (and did) misrepresent the chemicals as being less poisonous than they were [24]. When information was transferred with the chemical, there was no guarantee that information transferred with the chemical would be understood by the individuals utilizing the product. Barriers arose either because instructions were not written in the local language, or when the instructions were printed in a common language, but the applicator could not read [22].

10.2   Negotiating the Rotterdam Convention

States' interest in creating rules for importing hazardous chemicals significantly predates the formation of an International Negotiating Committee on the topic. Two international organizations, FAO and UNEP, issued guidelines for states on the topic. FAO published the International Code of Conduct for the Distribution and Use of Pesticides (Code of Conduct) in 1985. The document created a voluntary mechanism that states could use to establish public policy in developing countries that did not have the education, laws, regulations, or import controls necessary to prevent the misuse of pesticides. Such misuse directly led to the loss of life from pesticide poisoning and increased environmental damage. Zahedi [68] noted that developed countries did not believe they should ban hazardous chemicals and pesticides shipments to other developing countries as these states should decide for themselves.

Thus, the Code of Conduct sought to minimize the amount of environmental risk from using pesticides by encouraging states to create new public policies while also encouraging manufacturers to refrain from false advertising or other dubious business practices. For example, Article 9.5 recommended a notification from the exporting state to the importing state if the pesticide had been restricted or banned from use within the exporting state [15]. Importing countries would then need to create internal regulatory mechanisms for receiving and acting upon this information. This nascent regulatory scheme incorporated some, but not all, of the conditions necessary for a PIC scheme. In this case, notification to the importing country should occur, but informed consent that the importing state accepts the shipment is not mandated. Thus, the Code of Conduct only included notification, the first of the two steps needed for consent.

Hough [24] reviewed the impact of the Code of Conduct and determined that it made little direct impact on the exposure to pesticides in developing countries. He noted that many exporters ignored the scheme due to its voluntary nature. Equally important, manufacturers' export notifications sometimes arrived after the chemical shipment had been accepted at its notification, thereby circumventing the opportunity to decline a shipment. However, the Code of Conduct stimulated further diplomatic conversation on the need to govern the movement of pesticides across national boundaries.

The Code of Conduct contained other pertinent suggestions to limit the human health impacts of pesticides. One of these items included guidelines for pesticide manufacturers seeking to advertise their products. Article 11 of the Code of Conduct recommended that manufacturers should base their advertising claims on scientific facts, refrain from implying official acceptance of a chemical, or attempt to mislead a buyer about the chemical's effectiveness [15]. Additionally, companies should reiterate the need to use pesticides only as directed and to adhere to best safety practices when utilizing the product. While it is unclear the extent to which corporations' made a good faith effort to adhere to the guidelines, NGOs showed no difficulty criticizing companies for their perceived misbehaviors. Stevenson [40] wrote for an NGO that called out ICI for running an ad for the pesticide Paraquat in Malaysia in violation of the Code of Conduct.

The FAO recognized that the only way to truly eliminate the harm of pesticides would be to refrain from their use. Consequently, the Code of Conduct also sought to promote integrated pest management as a replacement for pesticide usage [15]. The integrated pest management system potentially allowed states to reduce the risk of exposure from pesticides to zero as this system frequently advocates for the elimination of pesticides. While this scheme may adhere more closely to the precautionary principle goals, it did not receive significant attention during this period.

Similarly, UNEP published the London Guidelines for the Exchange of Information on Chemicals in International Trade in 1987. Like the Code of Conduct before it, this document failed to incorporate a complete PIC regime due to resistance from the United States, the United Kingdom, and West Germany. Paarlberg [35] credited an NGO, the Pesticide Action Network, for successfully pressuring these reluctant states to negotiate a two-phased consent process instead of the one-phase notification process leading up to and during the 1987 Governing Council Session. Thus, the UNEP Governing Council, during its fourteenth session, passed Decision 14/27, directing Executive Director Tolba to strengthen the London Guidelines by incorporating the two-phased PIC process and reviewing the future need for a convention on this topic [45].

The London Guidelines and the Code of Conduct were modified in 1989 to reflect the PIC mechanism. However, Wirth [66] pointed out that while the London Guidelines, as revised in 1989, including the two-part PIC process, export countries did not have to abide by the importing country's decision. The requirement to refrain from transporting a pesticide that had been denied entry was not written into the London Guidelines.

The London Guidelines also attempted to address the information asymmetries between the developed country manufacturer and the developing country applicator by recommending that instructions for the use of the pesticide be communicated in the native language of the importing country rather than the manufacturing company [46]. Lowering a language barrier could potentially reduce the risk of human health exposure stemming from the inability to read the instructions.

States considered the issue at UNCED and included a recommendation in Agenda 21's Chapter 19.38(b) to create a legal-binding document that included participation and implementation of PIC by 2000 [55]. Thus, FAO in November 1994 authorized an INC to draft treaty text [16]. UNEP joined the efforts when its General Council approved joining the effort in Decision 18/12 in May 1995 with a goal to open the treaty for signature in 1997, three years before the Agenda 21 deadline [48].

Per international diplomatic tradition, the INC met multiple times to meet this deadline beginning with INC-1 on March 11–15, 1996, in Brussels, Belgium. At its first negotiating session, states identified a need to make the PIC processes from the Code of Conduct and the London Guidelines legally binding. Thus, delegates relied heavily on these documents and on further information obtained from the working groups on both the Code of Conduct and the London Guidelines that further debated items for inclusion in these two documents. Participants at INC-1 also discussed a need for the process to be compatible with current GATT and WTO trade requirements [52]. Points of contention also emerged, including resisting the temptation to move beyond the scope of the UNEP Governing Council mandate, selecting criteria for including chemicals, creating a pathway to add new chemicals, and providing financial and technical assistance. UNEP and FAO [52] noted states' objections to including a prohibition of the use or phase-outs of chemicals as part of this negotiating process. Kummer [31] provided slightly more detail on the discussion over the scope of the convention as she noted that the United States preferred to narrowly limit the conversation to a discussion of PIC, while the EU sought a broader-based treaty focused on chemical management.

INC-2 followed shortly after that from September 16–20, 1996, in Nairobi, Kenya. However, consensus proved more elusive than anticipated and states called for additional meetings to break the deadlock. The INC suffered the further indignity of needing to borrow money from UNEP's Environment Fund to pay for the second session. The combination of these two factors indicated that the treaty would likely have a small impact on the environment.

Additionally, the FAO Council issued further clarification on the treaty's scope during its 111th Session. It confirmed its commitment to provide logistical and technical support to the INC [17]. Similarly, UNEP's Governing Council also issued instructions to its Executive Director in Decision 19/13 during their meeting in

Nairobi from January 27–February 7, 1997 [49]. After the further clarification by the FAO Council and UNEP, the INC met nine additional times, refining the treaty text. While states initially hoped to complete the negotiations by the end of 1997, this did not occur. Additional negotiating sessions became necessary, leaving states, FAO, and UNEP covering the additional costs [18, 49].

The INC met three additional times from May 1997 to the Conference of the Plenipotentiaries in September 1998. During this time frame, delegates engaged in typical negotiating patterns that focused on identifying areas for consensus while leaving more contentious areas until the end. In this sense, the negotiating pattern followed a traditional pattern for environmental negotiations as the text for the Rotterdam Convention slowly emerged.

The Rotterdam Convention converted the voluntary PIC process into a mandatory process. The receiving country would be responsible for determining whether or not to accept the chemical. Thus, the treaty creates a procedure whereby states may inform others about their intent to receive a chemical. States wishing to ban imports of a chemical using this mechanism must simultaneously ban imports from all countries and enact a domestic ban on the chemical as well. This avoids the appearance of using the Rotterdam Convention as a disguise for a trade embargo [62]. Chemicals may be banned or severely restricted by their placement on Annex III or by one state notifying another state of its decision to restrict a chemical when not listed under Annex III. Importantly, Annex III is not a static list. The number of chemicals on the list changes as states agree to add additional chemicals. At the time of adoption, 27 chemicals were included within Annex III, five industrial chemicals and twenty-two pesticides. By January 2022, the list grew to 53 chemicals [54].

States may propose new chemicals to be included in the Annex III list to the Secretariat [56]. The Rotterdam Convention requires at least one proposal from each of two PIC regions out of the seven established utilizing the FAO classification scheme. After receiving the second notification, the treaty Secretariat authenticates the proposals and notifies other parties about the submission. The Secretariat also forwards the proposals to the Chemical Review Committee. The Chemical Review Committee creates a draft decision guidance document for the COP, which decides whether to add the chemical to Annex III. Should the COP choose to list the chemical in Annex III, the Secretariat will notify all parties of the decision. Similarly, states may also delist a chemical in Annex III utilizing a similar procedure.

Thus, the Rotterdam Convention created an information exchange where states who ban or severely restrict a chemical domestically must report the action to the Secretariat. This requirement allowed developing countries that might not otherwise have the domestic capabilities to test every chemical to rely on more advanced countries for scientific information and environmentally protective public policy development. By creating the information exchange mechanisms, diplomats hoped to avoid circumstances where chemicals were manufactured and transported to countries that could not successfully manage their use and disposal.

The Secretariat under the Rotterdam Convention plays a central role in managing information shared between countries. Therefore, it is unsurprising that states invited UNEP and FAO to continue serving as the Secretariat through the interim status.

These organizations had the experience of operating the Code of Conduct and London Guidelines. States did not finalize arrangements for a permanent secretariat and location during the Conference of Plenipotentiaries. Upon completion of the Conference, the Rotterdam Convention opened for signature in September 1988 and gathered 72 signatories, including the United States. It entered into force on February 24, 2004. States were encouraged to utilize a voluntary PIC mechanism in the interim. UNEP and FAO [53] noted that 167 states and the EU utilized the interim PIC process.

Scholars thoroughly criticized the Rotterdam Convention for a variety of failures dealing with the potential for loopholes, weaknesses in notification requirements both externally and internally, the absence of strong enforcement provisions, a failure to address underlying causes of pesticide misuse such as handler restrictions, labeling, packaging, or advertising, and the failure to incorporate a strict application of the precautionary principle [37, 68]. Zahedi [68] also expressed concern about the Rotterdam Convention's ability to add new chemicals over time as she believes the process for adding chemicals and pesticides to Annex III is overly burdensome. However, this critique may be unfounded as the list of chemicals and pesticides included in Annex III slowly grows. Additionally, states continue to meet and investigate other chemicals and pesticides for further restrictions.

The creation of the Rotterdam Convention did not force the retirement of the Code of Conduct of the London Guidelines. While the FAO has periodically updated the document on behalf of its member states, its guidelines continue to be needed as the Rotterdam Convention did not address all the pathways for pesticides to impact human health and the environment negatively. The treaty did not address standards for handler restrictions, labeling, packaging, or advertising. In practice, some complaints about pesticide misuse are likely to be handled through NGO pressure campaigns directly against corporations.

Zahedi [68] argued that NGOs should be brought into a formal arrangement under the Rotterdam Convention. Jansen and Dubois [27] disagreed by pointing out that the PIC process established in the Rotterdam Convention does not need NGO participation to function effectively. Attendance records from the interim Chemical Review Committee de facto settled the question of the role of NGOs during the interim period. NGOs participated as observers in the meeting, albeit in limited numbers [51]. Unsurprisingly, business and industry-related NGOs attended the meeting in higher numbers than environmental NGOs. This practice essentially left the role of NGOs unchanged as interested observers.

After the Conference of the Plenipotentiaries, states reconvened the INC to work on unresolved issues not included in the Rotterdam Convention. Five additional meetings were held in the INC framework. INC-11, the sixth and final meeting, met in a joint session with COP-1 of the Rotterdam Convention. An initial lack of funding for INC 6 forced states to focus on operational aspects of the Rotterdam Convention, such as the interim Chemical Review Committee, rather than addressing developing countries' concerns about new and additional funds to carry out their compliance obligations under the Rotterdam Convention [14]. Similarly, Geneva and Rome served as the interim Secretariat location, corresponding to offices for both UNEP and FAO during this period. States confirmed UNEP and FAO as the Secretariat with the location in Geneva and Rome at COP 1.

10.3 Negotiating the Stockholm POPs Convention

Persistent organic pollutants (POPs) are toxic chemicals that, due to their stability, do not easily break down in the natural environment [28]. Thus, these chemicals tend to last in the environment for extended periods [4]. Many POPs became prominent in chemical manufacturing, agriculture, and other industries after the second World War. As the chemical manufacturing sector expanded its manufacturing locations from developed countries to developing countries, states recognized the need for further actions regulating the production processes. In other words, it does not help the environment if a chemical manufacturer that has been banned from producing a POP moves to another country for the purposes of producing a POP then imports the banned chemical to the original locale. This situation potentially expands the risks as the chemical continues to be manufactured but increases the transportation distance with all its associated risks. Thus, an international regulatory scheme for controlling, reducing, and, where possible, eliminating POPs may significantly contribute to lowering environmental hazards.

POPs may be transferred through the atmosphere and water to locales long distances away from their release point. Further, POPs tend to bioaccumulate in fatty tissues of living species, moving from one animal to another throughout the food chain [11, 26]. POPs may have severe human health impacts at relatively low levels of exposure, primarily as endocrine disruptors that interfere with the hormone system [38]. These chemicals may also be linked to cancer in animals and humans [1]. POPs tend to concentrate in areas with colder weather, raising the bioaccumulation rate and risk profile, disproportionately impacting Indigenous Peoples in the Northern hemisphere [7].

While the Stockholm and Rotterdam Convention both share the concern of limiting exposure to hazardous chemicals and pesticides, these two treaties differ in their institutional origins. The two treaties can and do overlap in regulating the same chemicals. While the two Conventions undoubtedly share roots in common environmental risk exposures caused by hazardous chemicals, significant differences in the subject matter of their content nevertheless emerged. That is not to say that the two groups of negotiators considered the work to be entirely separate. Both scholars and practitioners recognized the interconnectivity between these two works [12, 37]. ENB [12] also expressed concern that negotiations under the Stockholm POPs might prove to be more difficult given the economic implications of a manufacturing ban. Thus, the Stockholm POPs Convention moved closer to a precautionary approach to chemical management in that the Convention eliminates environmental health impacts by eliminating the pathway of risk via a ban in production [67].

Efforts to ensure the sound management of chemicals to lower their individual and collective impact on human health and the environment began in the 1970s and

1980s. International activity intensified with the creation of the International Forum on Chemical Safety (IFCS) after states suggested its creation as part of Agenda 21's Chapter 19 [55]. ILO, UNEP, and WHO provided support for IFCS in 1994, with WHO providing facilities and necessary support. IFCS functioned as an advisory group consisting of public and private members to give policy guidance to relevant actors regarding the safe management of chemicals. An Ad Hoc Working Group on POPs formed by the IFCS identified the initial twelve chemicals disparagingly nicknamed the "dirty dozen" for their negative impact on the environment. The dirty dozen consisted of aldrin, chlordane, DDT, dieldrin, endrin, heptachlor, hexachlorobenzene, mirex, PCBs, polychlorinated dibenzodioxins (PCDDs), polychlorinated dibenzofurans (PCDFs), and toxaphene. Many of these chemicals work as a pesticide, with the exception of PCBs, a heat exchange fluid, and PCDDs and PCDFs that occur as a byproduct of combustion. For the purposes of regulation, the dirty dozen subdivides into two categories of chemicals, the so-called deliberate chemicals, and the byproducts.

In Decision 18/32 issued by the UNEP Governing Council as part of its eighteenth session in 1995, UNEP asked the IFCS and the Inter-Organization Programme for the Sound Management of Chemicals, working with the International Programme on Chemical Safety, to assess these twelve chemicals [47]. The IFCS conducted an initial assessment for the dirty dozen including their chemical characteristics, toxicity, the prevalence in the environment, and availability of substitutes before recommending that states negotiate a new international treaty to reduce risks of human health impacts from these chemicals to UNEP's Governing Council and the World Health Assembly, the operational body of the WHO [25]. Buccini [4] emphasized that the IFCS working group process ended the conversation about whether these chemicals should be controlled and allowed states to begin discussions about how to proceed with a new regulatory process.

UNEP's Governing Council in Decision 19/13C in 1997 and the World Health Assembly, in Resolution WHA50.13, agreed with the IFCS recommendation and authorized the creation of a negotiating committee on POPs [49]. Like the other conventions covered in this manuscript, this decision prompted the beginning of formal diplomatic negotiations through the creation of an INC. UNEP further estimated this group would meet five times from 1998 to 2000, with the Criteria Expert Group charged with determining scientific-based criteria for including chemicals in the future, meeting three times as well.

During the first INC meeting in Montreal, Canada, in 1998, the United Kingdom asked for and states made a deliberate decision to avoid overlapping with the pre-existing Basel Convention and the Rotterdam Convention while negotiating on the Stockholm POPs Convention [13, 61]. States identified areas of discussion early in the negotiating process, including the reduction and/or elimination of products and byproducts, the management and disposal of stockpiles, information exchange mechanisms, and implementation assistance [4, 13, 29]. While the issue of whether to expand the scope of chemicals beyond the initial twelve POPs occurred briefly at the beginning of the negotiating process, ENB [13] reported a near consensus opinion to leave the chemical list as is and to add a mechanism to review and add new chemicals to the treaty text.

For the POPs that are deliberately produced, states agreed to eliminate their production or use unless a specific exemption was registered, i.e., a country wished to use DDT for malaria control. States divided the ten chemicals manufactured as finished products into two lists, Annex A and Annex B. Nine of the ten so-called deliberate chemicals are listed in Annex A and are subject to a requirement to eliminate the production and use of these chemicals. Chemicals listed in Annex A are subject to elimination of production. States may register a specific exemption for a chemical. However, the specific exemption is only available to the state claiming the exemption. Specific exemptions may be claimed for producing or using a chemical for a limited time frame. The tenth chemical, DDT, is listed in Annex B and is subject to a restriction in production and use only. States made this distinction to allow for the use of chemicals that, while hazardous, may have an otherwise acceptable use, such as controlling diseases.

The remaining two chemicals, PCDD and PCDF, perhaps better known as dioxins and furans, occur as a by-product of combustion, such as waste incineration or power generation. For the POPs that occur as byproducts of other processes, elimination is not straightforward as this might require significant chemical manufacturing process changes. As a result, states agreed to various actions to manage policies and processes to minimize byproduct releases, including promoting the development of new manufacturing processes and technologies. In keeping with the idea of lowering human health risks from hazardous chemicals, the Stockholm POPs require states to adopt technological control measures. States agreed to utilize Best Available Techniques (BAT) and Best Environmental Practices that seek to minimize the Annex C chemicals emitted through a variety of process changes and operational practices as determined by the COP. However, Vanden Bilcke [61] noted that the requirement for BAT is ambiguous and weak.

The Stockholm POPs Convention encourages states to identify, manage, and destroy existing stockpiles of regulated chemicals, their wastes, or other items contaminated by a regulated chemical. Various storage, transportation, and destruction techniques may be allowable under the principle of environmentally sound management, with an end goal of eliminating the characteristics that make a chemical a POP [61]. Additionally, environmentally sound disposal may be allowable when destruction is not feasible. Further, site remediation can be undertaken at a parties' discretion.

Per Article 8, parties may submit a chemical for review to the Secretariat [50]. The Secretariat will review the proposal and determine whether the package is complete before advancing the information to the POPs Review Committee. The POPs Review Committee will then make the proposal and its evaluation available to all parties to contribute information so that the POPs Review Committee may create a draft risk profile. This document is circulated to parties for further discussion and to solicit information pertaining to the socioeconomic considerations of regulating the chemical. This new information will be utilized to create a draft risk management evaluation. The POPs Review Committee will make a recommendation to the COP who holds the final decision-making authority.

States designated UNEP as the Conference Secretariat in Article 20 and charged the Secretariat with maintaining an information exchange online, assisting parties with implementation, coordinating with other relevant bodies, and assisting with other duties as needed [50]. While a significant debate occurred during the negotiating process for this treaty, states nevertheless designated the GEF as the interim funding mechanism but left open the possibility of the COP selecting another organization to assume the responsibility after the Stockholm POPs Convention entered into force [61]. States confirmed the GEF as the funding mechanism at COP 2.

One of the unique features of the Stockholm POPs Convention involves its evaluation of environmental effectiveness under Article 16 [50]. States agreed to monitor the levels of POPs regulated under the treaty, their impacts on health and the environment, and to review this report four years after the date of entry into force. This review will not be a one-time review as states also agreed to design a mechanism for further review once the COP meets. Additionally, states intend to increase the number of chemicals regulated by the regime at a later date. However, states elected to forgo the framework-protocol convention, choosing instead to negotiate a substantive treaty. Karlaganis et al. [29] explained that states found it easier to ban identified chemicals than to design an open-ended process for adding unknown chemicals to the ban in the future. When the chemical is identified in advance, a state may make a rational decision whether to proceed by balancing environmental impacts with economic impacts. In the case where the future chemical is unknown, states hesitate as it is much more difficult to understand the impact of the future economic risks versus environmental impacts.

States completed the negotiations for the Stockholm POPs treaty in May 2001. It immediately opened the next day for signature. Representatives from ninety states signed the document on behalf of their country before departing the meeting. On May 17, 2004, the treaty entered into force after ratification by fifty countries. Like other conventions involving the regulation of hazardous chemicals, the United States has not ratified the treaty. This treaty's non-ratification is partly because the EPA does not have the legal authority necessary to carry out its obligations under the Toxic Substance Control Act or the Federal Insecticide, Fungicide, and Rodenticide Act [21].

The Basel Convention, the Rotterdam Convention, and the Stockholm POPs Convention collectively work together to regulate transnational movements of hazardous chemicals and their wastes. Given the interlinkages between these three conventions, states consolidated support mechanisms. UNEP and FAO coordinated to consolidate the administrative mechanisms into one secretariat, the Basel, Rotterdam, Stockholm (BRS) Secretariat, housed in Geneva, Switzerland. Similarly, states hold consolidated meetings for the three conventions.

10.4 The Hazardous Waste Regime

Returning to one of the fundamental concepts within international relations, the idea of a singular regime dealing with hazardous waste is on display in this chapter. The creation of the BRS Secretariat, along with numerous statements of diplomats and practitioners during the negotiations, indicated that actors recognized the interconnectedness of these international treaties. Thus, the hazardous waste regime could be considered successful in that an effective, well-run institution leads to creating and implementing regulations that improve environmental health.

Hidden behind the details of the successful outcomes of the Rotterdam and Stockholm Convention, however, is the closure of the IFCS. This short-lived institution also points to a facet not often examined within international environmental diplomacy. Institutions do not have to continue into perpetuity. The IFCS served a useful purpose as an ad hoc working group for a specific task, whereas empty institutions typically have no purpose beyond offering up statements for political momentum [10]. The IFCS made a concrete recommendation that turned into a treaty, thus removing it from the empty institution classification.

In addition to the classification of institution type, the hazardous waste regime also contains elements of both hard and soft law. One of the truisms of international law is that soft law hardens into treaty text and therefore becomes hard law. While it may be tempting to think that the creation of the Rotterdam Convention should have forced the end of the voluntary guidelines such as the Code of Conduct or the London Guidelines, this was not the case. Weiss [64] believed that soft law documents had value separate from whether they would become hard law over time for various reasons. They allowed for flexibility and for actors' expectations to emerge before investing in formal negotiating sessions. She noted that while lawyers may be more comfortable with specific legal deadlines that can be enforced, political scientists may be more inclined to focus on whether norms are established, or changes in behaviors occur. After the passage of these important treaties, the Code of Conduct and the London Guidelines did not fade into the background. The Code of Conduct continues to be maintained by FAO and incorporates elements of the internationally negotiated treaties into its guidelines. Additionally, the Code of Conduct, due to its non-binding nature, may make suggestions that would not be acceptable in a legally binding treaty. Thus, this document adds to the complexity of international environmental policy by establishing expectations for behaviors, even if the behaviors are not formally required.

In line with viewing the hazardous waste regulations through the lens of global environmental governance, Wirth [66] noted that the accountability of actors to states shifted from this hard compliance stance to a softer public domain due to the actions of NGOs. In other words, states no longer solely police the actions of their domestic entities or other states. Instead, NGOs utilize their own communications to target other actors through the use of negative publicity. Weiss [64] pointed out that compliance in international law may be better analyzed as a process rather than a fixed outcome in that states may be less likely to comply with a treaty immediately after its signature and ratification, but more likely to adhere to the treaty as it ages, as NGOs and others are more likely to demand compliance through informal mechanisms.

Karlanganis et al. [29] pointed out that he believes the success of the Stockholm POPs Convention depended not on the number of countries that adhere to banning the dirty dozen chemicals but rather on whether the expansion processes function as intended in the long term. While the number of chemicals regulated under these two regimes has not, perhaps, grown as quickly as some states and NGOs would have wished, the treaties have expanded their regulatory reach.

References

1. Alharbi OM et al (2018) Health and environmental effects of persistent organic pollutants. J Mol Liq 263:442–453
2. Ansbaugh N et al (2013) Agent orange as a risk factor for high-grade prostate cancer. Cancer 119(13):2399–2404
3. Blair A et al (2015) Pesticides and human health. Occup Environ Med 72(2):81–82
4. Buccini J (2003) The development of a global treaty on persistent organic pollutants (POPs). In: Fiedler H (ed) Persistent organic pollutants. Springer, Berlin, pp 13–30
5. Buès R et al (2004) Assessing the environmental impacts of pesticides used on processing tomato crops. Agr Ecosyst Environ 102(2):155–162
6. Carson R (1962) Silent spring. Houghton Mifflin, Boston
7. Colborn T et al (1996) Our stolen future: are we threatening our fertility, intelligence and survival? A scientific detective story. EP Dutton, New York
8. Colopy J (1995) Poisoning the developing world: the exportation of unregistered and severely restricted pesticides from the United States. UCLA J Environ Law Policy 13(2):167–223
9. Deziel NC et al (2015) A review of nonoccupational pathways for pesticide exposure in women living in agricultural areas. Environ Health Persp 123(6):515–524
10. Dimitrov RS (2020) Empty institutions in global environmental politics. Int Stud Rev 22(3):626–650
11. El-Shahawi MS et al (2010) An overview on the accumulation, distribution, transformations, toxicity and analytical methods for the monitoring of persistent organic pollutants. Talanta 80(5):1587–1597
12. ENB (1998a) Report of the fifth session of the INC for an international legally binding instrument for the application of the prior informed consent procedure for certain hazardous chemicals and pesticides in international trade: 9–14 March 1998. 15(4):1–11
13. ENB (1998b) Report of the first session of the INC for an international legally binding instrument for implementing international action on certain persistent organic pollutants (POPS): 29 June–3 July 1998. 15(10):1–10
14. ENB (1999) Report of the sixth session of the INC for an international legally binding instrument for the application of the prior informed consent procedure for certain hazardous chemicals and pesticides in international trade: 12–16 July 1999. 15(20):1–10
15. FAO (1985) International code of conduct for the distribution and use of pesticides. FAO, Rome
16. FAO (1994) Report of the council of FAO. CL/107/REP
17. FAO (1996) Report of the council of FAO hundred and eleventh session Rome, 1–10 Oct 1996. CL 111/REP
18. FAO (1997) Report of the council of FAO hundred and thirteenth session Rome, 4–6 Nov 1997. CL 112/REP
19. Freire C et al (2013) Long-term exposure to organochlorine pesticides and thyroid status in adults in a heavily contaminated area in Brazil. Environ Res 127:7–15

20. Fuhrimann S et al (2022) Pesticide research on environmental and human exposure and risks in Sub-Saharan Africa: a systematic literature review. Int J Environ Res Pu 19(1):259–276
21. Hagen PE, Walls MP (2005) The Stockholm convention on persistent organic pollutants. Nat Resour Environ 19(4):49–52
22. Halter F (1987) Regulating information exchange and international trade in pesticides and other toxic substances to meet the needs of developing countries. Columbia J Environ Law 12(1):1–38
23. Hilson C (2005) Information disclosure and the regulation of traded product risks. J Environ Law 17(3):305–322
24. Hough P (1996) Stemming the flow of poison the role of UNEP and the FAO in regulating the international trade in pesticides. Inter Relat 13(1):69–79
25. IFCS (1996) Final report of the Ad Hoc working group on persistent organic pollutants meeting. IFCS/WG.POPs/REPORT.1
26. Jamieson AJ et al (2017) Bioaccumulation of persistent organic pollutants in the deepest ocean fauna. Nat Ecol Evol 1(3):1–4
27. Jansen K, Dubois M (2014) Global pesticide governance by disclosure: prior informed consent and the Rotterdam convention. In: Gupta A, Mason M (eds) Transparency in global environmental governance: critical perspectives. MIT Press, Cambridge, pp 107–132
28. Jones KC, De Voogt P (1999) Persistent organic pollutants (POPs): state of the science. Environ Pollut 100(1–3):209–221
29. Karlaganis G et al (2001) The elaboration of the 'Stockholm convention' on persistent organic pollutants (POPs): a negotiation process fraught with obstacles and opportunities. Environ Sci Pollut Res 8(3):216–221
30. Kuhnlein HV, Chan HM (2000) Environment and contaminants in traditional food systems of northern indigenous peoples. Annu Rev Nutr 20(1):595–626
31. Kummer K (1999) Prior informed consent for chemicals in international trade: the 1998 Rotterdam convention. Rev Eur Comp Inter Environ Law 8(3):323–330
32. Lu C et al (2000) Pesticide exposure of children in an agricultural community: evidence of household proximity to farmland and take home exposure pathways. Environ Res 84(3):290–302
33. Macharia IN et al (2009) Potential environmental impacts of pesticides use in the vegetable sub-sector in Kenya. Afr J Hortic Sci 2:138–151
34. Mazur CS et al (2015) P-glycoprotein inhibition by the agricultural pesticide propiconazole and its hydroxylated metabolites: implications for pesticide–drug interactions. Toxicol Lett 232(1):37–45
35. Paarlberg (1992) Managing pesticide use in developing countries. Environ: Sci Policy Sustain Dev 34(4):17
36. Rocha GM, Grisolia CK (2019) Why pesticides with mutagenic, carcinogenic and reproductive risks are registered in Brazil. Dev World Bioeth 19(3):148–154
37. Ross J (1999) Legally binding prior informed consent. Colo J Inter Environ Law Policy 10(2):499–530
38. Ruzzin J et al (2012) Reconsidering metabolic diseases: the impacts of persistent organic pollutants. Atherosclerosis 224(1):1–3
39. Sabarwal A et al (2018) Hazardous effects of chemical pesticides on human health–cancer and other associated disorders. Environ Toxicol Phar 63:103–114
40. Stevenson P (1993) FAO code violations in advertising. Glob Pestic Campaigner 3(4):3
41. Swaminathan M (1982) Politics of pesticides. Nat 296(5857):521–522
42. Tait J, Bruce A (2001) Globalisation and transboundary risk regulation: pesticides and genetically modified crops. Health Risk Soc 3(1):99–112
43. Thongprakaisang S et al (2013) Glyphosate induces human breast cancer cells growth via estrogen receptors. Food Chem Toxicol 59:129–136
44. UNEP (1982) UNEP: report of the governing council (session of a special character and 10th session, 1982). A/37/25
45. UNEP (1987) Report of the governing council on the work of its fourteenth session. A/42/25

46. UNEP (1989) London guidelines for the exchange of information on chemicals in international trade amended 1989—decision 15/30 of the Governing Council of UNEP of May 25 1989. UNEP, Nairobi
47. UNEP (1995a) Persistent organic pollutants. UNEP/GC/DEC/18/32
48. UNEP (1995b) Report of the governing council on the work of its eighteenth session. UNEP/GC/18/40
49. UNEP (1997) Proceedings of the governing council at its nineteenth session. UNEP/GC.19/34
50. UNEP (2001) Stockholm convention on persistent organic pollutants. UNEP/POPS/CONF/2
51. UNEP and FAO (2000) Report of the interim chemical review committee on the work of its first session. UNEP/FAO/PIC/ICRC.1/6
52. UNEP and FAO (1996) Intergovernmental negotiating committee for an internationally legally binding instrument for the application of the prior informed consent procedure for certain hazardous chemicals and pesticides in international trade. UNEP/FAO/PIC/INC.1/10
53. UNEP and FAO (2003) Report of the intergovernmental negotiating committee for an international legally binding instrument for the application of the prior informed consent procedure for certain hazardous chemicals and pesticides in international trade on the work of its tenth session. UNEP/FAO/PIC/INC.10/24
54. UNEP and FAO (2022) Annex III chemicals. http://www.pic.int/TheConvention/Chemicals/AnnexIIIChemicals/tabid/1132/language/en-US/Default.aspx. 25 Jan 2022
55. UN (1993) Agenda 21: programme of action for sustainable development, Rio declaration on environment and development, statement of forest principles: the final text of agreements negotiated by governments at the United Nations Conference on Environment and Development (UNCED), 3–14 June 1992, Rio de Janeiro, Brazil. UN, New York
56. UN (1998) Rotterdam convention on the prior informed consent procedure for certain hazardous chemicals and pesticides in international trade. UNEP/CHEMICALS/2000/2
57. Uram C (1990) International regulation of the sale and use of pesticides. Northwestern J Inter Law Bus 10(3):460–478
58. Van den Berg H (2009) Global status of DDT and its alternatives for use in vector control to prevent disease. Environ Health Persp 117(11):1656–1663
59. Van den Berg H et al (2017) Global trends in the production and use of DDT for control of malaria and other vector-borne diseases. Malaria J 16(1):1–8
60. Van der Werf HM (1996) Assessing the impact of pesticides on the environment. Agr Ecosyst Environ 60(2–3):81–96
61. Vanden Bilcke C (2002) The Stockholm convention on persistent organic pollutants. Rev Eur Comp Inter Environ Law 11(3):328–342
62. VanDorn HM (1999) The Rotterdam convention. Colo J Intl Envtl L Poly 10(1998 Yearbook):281–290
63. Weir D, Schapiro M (1981) Circle of poison: pesticides and people in a hungry world. Food First Books, Oakland
64. Weiss EB (1999) Understanding compliance with international environmental agreements: the baker's dozen myths. Univ Richmond Law Rev 32(5):1555–1590
65. WHO (2021) World malaria report. WHO, Geneva
66. Wirth DA (1990) Remarks by David A. Wirth. In: Proceedings of the ASIL annual meeting, vol 84 Cambridge University Press, Cambridge, pp 145–151
67. Yoder AJ (2003) Lessons from Stockholm: evaluating the global convention on persistent organic pollutants. Indiana J Glob Leg Stud 10(2):113–156
68. Zahedi NS (1999) Implementing the Rotterdam convention: the challenges of transforming aspirational goals into effective controls on hazardous pesticide exports to developing countries. Georget Inter Environ Law Rev 11(3):707–740

Chapter 11
Implementing Goals and Targets for Sustainability

Abstract At the beginning of the twenty-first century, Secretary-General Kofi Annan recommitted the United Nations to eliminating poverty around the world. To do so, the United Nations and other relevant international organizations such as the World Bank Group, and the Organization for Economic Co-operation and Development, would work together to articulate the Millennium Development Goals for 2000–2015. In establishing a set of goals for all actors of society to achieve together, states broadened the tools utilized to cooperate effectively. Scholars previously noted the onset of global environmental governance, a term utilized to reflect those actions taken away from state-dominated diplomatic processes also add to the patterns of behaviors and expectations that govern the international arena.

Keywords Millennium development goals · Poverty eradication · Global governance · World Bank Group · Sustainable development · OECD

On September 8, 2000, UN Secretary-General Kofi Annan (Ghana) announced the MDGs as part of the Millennium Declaration. A Nobel Peace Prize recipient, two-term Secretary-General Annan represented the best of the ideals that created and continue to sustain the modern agenda of the UN. Equally noteworthy, Annan subtly shifted the role of the UN from a state-centric organization to one focused on both people and states. While Annan undoubtedly expected states to take care of their citizens, he recognized that there are times and places where this does not happen for various reasons, including lack of resources to do so. Accordingly, he positioned the UN to serve as the key hub for an increasingly dense organization of groups, including states, international organizations, businesses and industries, and NGOs focused on promoting the quality of life for all citizens [35].

The UN sought and continues to seek to promote the quality of life for all by promoting economic prosperity, better access to health services, education, sanitation, and human rights through the MDGs. These goals and targets encourage states to look after their citizens by collecting data and publishing the statistical averages of developing countries in the South. Additionally, this data can also be used to benchmark the developing countries' status and progress against the developed countries in the North. Scholars may refer to this new approach to global cooperation as governance through goal setting [8, 22].

© The Author(s), under exclusive license to Springer Nature Switzerland AG 2022   175
A. Egelston, *Worth Saving*, AESS Interdisciplinary Environmental Studies and Sciences Series, https://doi.org/10.1007/978-3-031-06990-1_11

Ironically, the origins of the UN's main programs for development had its roots outside of the UN system. The initial conceptualization arose from the OECD, an independent international organization founded in 1961 to promote global trade. Today, the OECD contains a scant 37 members (compared to the UN's 193 members). Further, the OECD members represent North America and Europe with limited participation from South America and the Asia Pacific and no participation from Africa. However, these 37 member countries represent approximately 80% of world trade and investment [34]. Consequently, OECD member countries have significant sway in any international fora.

The notion of goals and targets as an indicator of economic wealth is certainly not new. Economists have used important measures of wealth such as Gross Domestic Product and the Gross National Product to measure the economic wealth of a country. Today, one of the primary poverty indicators includes an estimate of how many people live on a fixed dollar amount. For example, the World Bank Group [47] estimated that in 2017, 689 million people lived on less than $1.90 per day (approximately $693.50 per year). Many of these global citizens are concentrated in Sub-Saharan Africa, including Nigeria, the Democratic Republic of Congo, Tanzania, Ethiopia, and Madagascar [47]. In contrast, the United States gross domestic product per capita (wealth per person) is $63,051 for 2020 [26]. This estimate places the average wealth (but not the average income) of a United States resident seventh in the world behind other countries such as Luxembourg, Singapore, Qatar, Ireland, Switzerland, and Norway. These startling differences indicate the vast inequalities in people's ability to provide for their basic needs, including food, water, sanitation, shelter, and clothing. Further, these differences also represent differences in access to natural resources, ability to secure a comfortable lifestyle, susceptibility to disease, education, social equality, or endangerment due to a poor environment.

Annan's visionary call represented a vast departure for the UN from the hallowed halls of traditional economic development. The UN's focus on sustainability recognized that the natural environment must be incorporated into all aspects of human society [4]. Thus, the UN sought to focus its own resources on those who truly have neither the goods, the education, nor the access to global power structures to change their present impoverished condition. It also recognized the role and resources focused on poverty reduction and basic needs from non-state actors [42].

This chapter begins by presenting information about the various international organizations involved in articulating the MDGs. Unlike other treaties, the MDGs' conceptualization originated outside the UN system but not outside the so-called society of states. This review consists of a brief overview of international organizations' role in managing the global economy on behalf of the major states and concludes by discussing international organizations' role within the global political system. The second section presents a brief history of the origins of international development goals with its roots in the OECD through its transformation into the MDGs. Thus, this section looks inside the admittedly complex UN system and the UN Secretary-General's office, a prominent voice in creating the MDGs. This section also contains an overview of the content of the eight MDGs along with their targets. The third section returns to the UN for the Millennium Summit, a three-day event from September 6–8, 2000, culminating in the Millennium Declaration signing. Rather than reviewing all the myriad of events, workshops, and conferences that contributed to the implementation of the MDGs, the section highlights results and lessons learned in the process of carrying out the MDGs. The fourth section concludes by reviewing the changes in the international system brought about by the MDGs. Standing in direct contrast to many of the environmental episodes presented throughout this book, the MDGs differ in that this episode does not involve a singular problem or an international treaty with legal obligations, but rather seeks to alter fundamental socioeconomic relationships. Instead, the MDGs attempt to use the UN's moral authority to focus energy and resources on the persistent problem of poverty.

11.1 International Organizations

International organizations may be thought of as an entity established through an international treaty, agreement, or compact, that includes at least three states as members, and that supports activities in these states [23]. One frustratingly simple description of an international organization is that it exists at the bequest of states that voluntarily meet to temper conflict and promote cooperation. This definition may be an apt description of part of the role and function of an international organization in today's society. But this description does not capture the full range of tasks, power bases, or authority of an international organization. Academic scholarship increasingly views these entities as important actors with the ability to shape events and agreements, coordinate action, and secure effective international programs [31]. This is perhaps not surprising given that an international organization's actions, authority, power, and legitimacy have changed dramatically in the last century, especially in the post-Cold War era.

International organizations occur above the state at the subregional, regional, and global levels. Prominent global organizations, such as the UN, the OECD, the World Bank Group, and the IMF, all were created by treaties negotiated between states that gave the organization a specific legal character. While the tasks, forms, and functions of international organizations vary, international organizations nevertheless provide order and continuity to the international system. Barnett and Finnemore [5] point out that international organizations tended to develop their own moral authority in order to accomplish their assigned tasks. Thus, while international organizations may act on behalf of a wide range of states from one to all, international organizations may act on their own behalf or the behalf of others, as well.

As noted in Chap. 1, the origins of our current global system had their roots in the immediate aftermath of World War II. As the war wound down, the Allied victors turned their attention to two key tasks—rebuilding war-torn Europe and ensuring that a third world war did not occur. While it is relatively easy to declare Europe as recovered from World War II, the task of ensuring a world free from war truly never ends. Each of the three international organizations reviewed below came into existence to help manage international trade in goods and services. Additionally, many of these groups' functions have shifted throughout their history.

The OECD originally operated as the Organization for European Economic Cooperation. Its key task was to allow European countries to meet to discuss their need for aid from the United States to Europe under the Marshall Plan. This international aid program financed rebuilding Western Europe. The group was specifically charged with providing input on the allocation and distribution of United States aid to rebuild European agricultural and industrial manufacturing and rebuild critical infrastructures such as cities, roads, and railroads. As Europe emerged from the devastation caused by World War II, the Organization for European Economic Cooperation transitioned into the OECD in 1961 with its headquarters in Paris, France. The OECD promotes international trade and economic prosperity among its member countries. The OECD provides a meeting place for the financial elite to harmonize further and reconcile its economic policies ranging from taxes and tariffs to giving insight on the economic well-being of countries by publishing a statistical analysis of each country's financial health. The OECD also ensures that markets remain open and are free from unfair trade practices, including bribery and corruption.

While the OECD began its life as a way for European countries to express their needs for aid, the International Bank for Reconstruction and Development (IBRD) gave loans to countries to help with their reconstruction activities. Today, this organization is technically the "lending arm" within the larger World Bank. Initially founded in 1944 and headquartered in Washington, DC, this international organization initially focused on Western Europe, although developing countries expressed an interest in receiving loans, or applied for loans as well. IBRD arranged loans for items that private banks historically would not fund such as roads or bridges or other large infrastructure projects intended to stimulate countries' domestic economies. The World Bank Group operated and continues to operate using contributions from developed countries and combines this funding with the repayments from previous loans. This financial flow is then loaned to countries as capital on various projects.

As new states continued to be created during the 1960s and 1970s, states and IBRD recognized the need to continue to work to improve the condition of humanity. Thus, additional financial institutions within the World Bank Group came into existence. For example, the International Financing Corporation began in 1961, focusing on lending to private companies within developing countries. Consequently, the creation of the International Financing Corporation expanded the number of projects as well as the types of projects receiving funding.

Under the tenure of World Bank President Robert McNamara, the World Bank shifted its focus to financing projects that met the "basic needs" of middle-income countries. In exchange for lending the money, states agreed to make policy changes inside the country in addition to repayment of the loan. However, the World Bank Group's loan assistance frequently came with mandated requirements to allow free market access, sometimes with devastating impacts on those least able to bear the costs. In the late 1980s, the World Bank Group was openly blamed for impoverishing the very people this organization was intended to help [46].

While the World Bank Group has been moderately successful in helping countries improve their development status, it has a mixed track record. It is sometimes seen as a symbol of Western economic dominance [25, 36]. Calls for reform at the World Bank Group occur on a semi-regular basis, and the World Bank Group has been restructured multiple times. Nevertheless, the sheer volume of capital flowing through the World Bank Group gives it significant political power in shaping international policy, especially on sustainable development and poverty reduction issues.

The International Monetary Fund (IMF) also assisted in creating the MDGs. Like the other two international organizations mentioned above, the IMF has its roots in the post-World War II global reconstruction architecture. While the World Bank Group existed to finance infrastructure projects, the IMF's purpose included managing monetary cooperation. In other words, the IMF worked and continues to ensure that trade in currencies occurs in a timely and appropriate fashion as a means of facilitating the buying and selling of other goods and services.

Access to hard currencies potentially impacts a county's ability to buy goods and services on behalf of its citizenry. Various currencies may have dramatically different supplies and demands. For example, more people may demand a United States dollar because it is more widely accepted versus the Turkish lira or the Russian ruble. Contracts to purchase goods and services that cross national boundaries will need to specify the payment currency. Both the buyer and the seller will then estimate the currency exchange rate and how much the exchange rate may change over time as part of their due diligence in signing the contract. With respect to development, countries with a less popular currency may have to pay more to exchange their currency into a more popular one, thus limiting their ability to trade in the future. Thus, the ability to exchange currencies at a fixed exchange rate significantly strengthens the ability of countries to engage in international trade.

Both scholars and students profess an interest in how these vitally important organizations impact global society. However, there is no singularly accepted theoretical framework for analyzing the actions of these (or other) international organizations. Hurd [20] points out that international organizations fulfill three major functions simultaneously. First, international organizations may be actors in their own right. As we shall see in this chapter, the UN wields considerable power in establishing and directing global actions, especially on environmental and developmental issues. Second, international organizations may be pawns in the hands of other powerful actors, such as states. One of the World Bank Group critiques is its propensity to act on behalf of the United States. In practice, the United States government controls who becomes the president of the World Bank. As a result, policy choices tend to reflect what the donor countries want rather than what the borrower countries desire. Third, international organizations provide meeting space for states to discuss issues of importance. From the early origins of international environmental diplomatic history, the UN remains one of the key locales for meetings. These meetings are not typically held in New York City at the UN headquarters but rather rotate around the world with diplomats traveling extensively throughout their time in office.

11.2   The Draft Emerges

The key idea from the summary above is that a significant amount of international activity focuses on the needs of developing countries to create economic growth alongside reducing poverty. Technocrats inside multiple organizations including the World Bank, the OECD, and the UN worked both individually and cooperatively on finding new institutional solutions to encourage states and non-state actors to contribute to this goal. Recalling that the twentieth century closed with the promise of a "peace dividend" from the end of the Cold War, the international system appeared poised to enter a new era of global cooperation. The cessation of hostilities between East and West created an opportunity to refocus attention on human rights and development issues. However, refocusing did not occur. Instead, the major Western powers that ostensibly won the Cold War took a sharp turn toward the conservative side of the political spectrum and focused instead on balancing their budgets and domestic priorities. As this reality set in, the UN system and many other international development organizations were facing the very real possibility of severe budget cuts and downsizing. Similarly, the developing South was at risk of losing its ability to access funding to continue to develop economically.

Scholars pointed out that goals and targets were strategically embedded in other conferences throughout the international agenda [9, 19, 32, 40]. Thus, the idea of global goal setting was not something new but rather an accepted practice, albeit one that might well be ignored rather than funded and implemented. Setting goals and targets for development assistance would thus serve two primary purposes. First, it sets out a justification to continue the cash flows to developing countries through the existing international structure [19]. Second, the goals and targets would appeal to conservative factions within developed countries who, theoretically speaking, would be able to direct financial flows where they make the most significant gains in the target being measured [11].

In many ways, the idea of an overarching set of goals for development reflected contemporary trends within global politics during this decade. This idea of assembling a set of goals and targets to measure development emerged from quiet conversations in privileged hallways between the developed countries' international elite spanning from the OECD to the World Bank to the UN. The Development Assistance Committee, a subcommittee within the OECD, formed an informal workgroup known as the Groupe de Réflexion that expressed concern about future levels of funding development assistance [9]. Ideas for inclusion in a draft statement were drawn from previous UN mega-conferences in the past ten years, including such topics as poverty reduction, children's health, gender equality, and economic development [19].

The final document, *Shaping the 21st Century: The Contribution of Development Cooperation*, debuted during the OECD High-Level Meeting on May 6–7, 1996 [33]. While the document received a warm welcome in Europe, it received scant attention in other parts of the world, notably the United States. The document contained just three categories—economic well-being, social development, and environmental sustainability and regeneration. Its goals (there were no targets) included reducing extreme poverty, providing access to universal primary education, reducing mortality rates during and after childbirth, providing access to reproductive health services, and implementing sustainable development.

Once released, technical staff at other international organizations reviewed the document, including the UN, the World Bank Group, and the IMF. Hulme [19] credits Claire Short, Secretary of State of International Development for the United Kingdom, for promoting these goals during her tenure in office beginning in 1997. Within the next three years, these groups issued reports that began the process of delineating a preliminary set of goals and targets for international development.

The IMF, OECD, World Bank Group, and the UN collaborated on the publication of *A Better World for All: Progress toward the International Development Goals* in June of 2000 [9,19]. The goals selected for inclusion in this report are similar to the earlier 1996 *Shaping the 21st Century* document. The fact that these items were included again as development goals emphasize the dire consequences of poverty, children's lack of education and early mortality rates, women's restricted access to reproductive health services, and high mortality rates. Environmental quality was also selected for inclusion in this report. However, Secretary-General Annan's personal interest and efforts to promote developmental targets would quickly overshadow these reports.

The next draft document emerged from obscurity within the hallways of the UN headquarters in New York. Jones [21] reported that the MDGs emerged from a technocratic process rather than diplomatic negotiations involving compromises. The writing process took place out of the UN Secretary-General's office, with UN Assistant Secretary-General John Ruggie overseeing what would become *We The Peoples, The Role of the United Nations* in the twenty-first century [3]. The UN released the report in April 2000. It contributed to the refinement of the UN mission by articulating a people-focused vision for the UN. However, the UN chose to work through the state and leave the care of individuals to domestic policy. The role of individuals is left to domestic policy, and, in practice, a wide variety of views persist, ranging from the philosophy that the national government serves the people to the opposite view, the people serve the state.

The realization that the UN was actively seeking to define development goals ignited a bevy of activity among actors of all types. All actors recognized that the new report would define the global agenda. Inserting a favored action item into a declaration of this magnitude all but ensured dominance on the international diplomatic agenda for years, if not decades to come. Consequently, the writers of what would become the Millennium Declaration had an unenviable task. Of the many hundreds of suggestions and many thousands of potential groups and organizations to offend, this group picked the winners and the losers by their silence.

Technocrats, individuals with technical training that de facto manage society within a governmental entity, combed through existing pledges, commitments, goals, and targets that states accepted previously and simplified the final goals into eight. Each of the eight primary goals is further supported by targets, designated by a letter (i.e., Goal 3, Target A, or simply 3A). These vary in nature between quantitative and non-quantitative targets.

The first goal, eradicating poverty and hunger, asks for assistance in halving the number of people living in extreme poverty, less than $1.25 per day, and halving the number of hungry people. Additionally, the goal seeks to find employment for all. The second goal, achieving universal primary education, encouraged greater access to primary education. In doing so, it de facto argues for a reduction to child labor as children cannot work and attend school simultaneously. While the third goal highlights a need for social equality for young girls and women, its sole target overlaps with goal 2. Target 3.A called only for improved access to education for women by advocating for equal access to primary and secondary education as men.

The fourth and fifth goals focused on reducing the mortality rates of the vulnerable. Goal four focused on reducing the mortality of children under five by two-thirds. Goal five sought to improve the maternal mortality ratio, an indicator of the number of deaths during childbirth. The ratio is calculated by dividing the number of deaths during childbirth by 100,000 live births. This goal also included access to reproductive health services such as an experienced provider and an established clinic or midwife for assistance during delivery.

The sixth goal, sought to limit HIV/AIDS, malaria, and other diseases. The HIV/AIDS pandemic began in 1981 with reports of the first cases. HIV attacks the body's auto-immune response system, rendering it essentially inoperable and unable to fight off other diseases. While HIV/AIDS has no cure and is always fatal, drug therapies have advanced so that individuals with HIV may lead a significantly longer life before the acquired immunodeficiency syndrome (AIDS) phase of the disease begins.

The epicenter of this outbreak was and continues to be in Sub-Saharan Africa, where HIV/AIDS is a leading cause of morbidity [12]. The prevalence of HIV/AIDS is of great concern for the deeply impacted countries and the international community. The Sub-Saharan Africa region is also the locale for many underdeveloped countries with the least access to medical knowledge, pharmaceuticals, and finances. The global community continues to be concerned that HIV/AIDS in this area could spread and therefore threaten the health of individuals around the world. In addition to HIV/AIDS, the MDGs also specifically targeted malaria, a severe disease transmitted by mosquitos and prevalent in Africa.

The seventh goal, ensuring environmental sustainability, focused on the natural and built environment. Focus areas under this goal included wise use of natural resources for development, slowing biodiversity, and reducing the amount of ozone-depleting substances and greenhouse gases in the atmosphere. Concerns about the built environment included access to safe drinking water and sanitation. This area also acknowledged a problem with habitat. It sought to limit the number of people living in slums with substandard housing that may be constructed with deteriorated building materials and lack electricity, water hookups, or sewer systems.

The eighth goal, developing a global partnership for sustainable development, sought to broaden the number of participants actively seeking sustainability. Despite the wording of this goal that implies that the UN is seeking alliances outside of the ranks of states, this goal focuses instead on systemic issues, primarily involving the relationships between states. One of the key metrics in this area is official development assistance (ODA), an economic indicator of the financial gift from one country to another. The MDGs prioritized the overall amount of ODA sent to the LDCs that needed the most assistance. The countries that need the most help are often land-locked developing countries hampered in their ability to trade by the absence of access to an ocean or sea and the small island developing states with limited natural resources to use in the development process. In addition to ODA, Goal 8 also looked at the issue of international indebtedness, access to affordable drugs and pharmaceuticals, and adaptation to emerging technologies, especially the areas of information and communications.

The MDGs quickly and deeply penetrated the discourse, if not the consciousness of the UN, as many programs and secretariats redefined their work priorities according to the MDGs. The Secretariat of the Basel Convention [41] made a typical claim demonstrating the power of this idea when it asserted that "The Basel Convention plays a decisive role in achieving the MDGs—poverty reduction, reducing child mortality, improving maternal health, ensure environmental sustainability." However, this did not necessarily mean that the work each group performed changed dramatically. The Basel Convention, for example, will continue to focus on waste minimization, but the reasons for doing so will come from the MDGs.

The UN Secretary-General charged UNDP with collecting, analyzing, and disseminating data on each country's progress in meeting the MDGs. Each country does so by submitting an MDG Country Report that indicates both successes and failures in meeting the MDGs. The UN system viewed the expansion of development data as a vital outcome of the MDGs.

Given that, better data creates opportunities to focus attention on understanding the root cause of problems and better decision-making to solve those problems, governmental entities benefitted from the MDG goals and targets [44]. Sachs [39] pointed out that the MDGs' success is due, in part, because the MDGs are designed to allow non-state actor participation. In other words, any action that is taken anywhere in the world is captured within the MDGs as the data set is a collection of each country's aggregate data. The movement toward development may occur because of state action. Still, it may also happen because a non-state actor, a religious organization, an NGO, a business, or some combination of these groups took action. Furthermore, the MDG design does not allow an opt-out option. Even if the organization did not align itself with the MDGs, any benefit would nevertheless be captured in the aggregate data analysis.

Complaints about the MDGs emerged quickly along two dimensions. The first involved criticisms of the process used to develop the MDGs. In short, the technocratic drafting process led to questions about the legitimacy of the MDGs. For example, critics of the MDGs point out that they are themselves a facet of Northern domination as the MDGs were developed predominantly by the United States, Europe, and Japan without the input of the countries that would be required to take concrete actions [1]. Richard et al. [37] noted that very few developing countries were consulted in this process.

Jones [21] disagrees and will later argue that the technocratic drafting of the MDGs should be seen as a positive facet rather than a negative by pointing out that consensus is easier to achieve in small numbers than larger ones. Sachs [39] agreed with Jones by pointing out that the MDGs cost very little time or effort to negotiate, given the document's origins.

The critiques about the process of establishing the MDGs do not negate the comments about the goals themselves. The second dimension involved the content of the MDGs. Significant frustration occurred as many scholars and observers perceived that the MDGs left out critical components of development [15, 18, 27, 45]. Specific items of concern included the watering down of human rights, democracy, women's rights, and access to water.

11.3 All Important Implementation

The UN Millennial Campaign launched officially in 2002. The immediate onslaught of activity indicated that the MDGs captured the global public's imagination from the onset. At once, a rallying cry for action at the local level and a plea for further ODA from developed countries, the MDGs consistently motivated renewed action to improve the lives of millions. The MDGs represent a success in that they shaped the discourse for the entire UN system and beyond. Sachs [39] credits the simplicity of the MDGs for their durability. He lauds that "eight simple goals that fitted well into one poster!" allowed media, organizations, and individuals worldwide to understand the moral imperative to achieve these goals for all.

Richard et al. [37] provided insight into the initial strategies that UN agencies utilized to achieve these goals. Initially, countries were encouraged to pursue "quick wins" but later shifted focus to pursue "high impact strategies" that could deliver results on multiple goals, targets, and indicators. This strategic planning process focused on a combination of quick wins, followed by high-impact strategies.

This strategic implementation raised additional concerns about equity between the developing countries. In a data-driven world, donors could focus on countries most likely to meet or exceed quantitative targets. An action like this could direct funding away from the LDCs toward the developing countries close to meeting the MDGs [13]. While not focusing on geography per se, Haines and Cassels [17] raise an interesting question about the wisdom of increasing financial flows in the least developing countries that may not have the infrastructure needed to turn the investments into an improved health care system. In essence, Haines and Cassels point out the need for increased capacity building in social systems to more effectively utilize donor funds.

Officially, UNDP [43] proclaimed the MDGs a success. However, a more nuanced look at the goals and targets suggests that very few countries have progressed the MDGs. Sachs [39] reported that much of the progress made on Goal 1, Eradicating extreme poverty and hunger, may be attributed to the gains made in China, without respect to any other country. The World Bank Group [48] agreed and pointed out that much of Africa and South Africa did not meet the target for halving the number of people living in extreme poverty or halving the number of hungry people.

Rosenbaum [38] reported that the distinguishing feature on whether countries met the MDGs is their status within the developing countries. Developing countries with higher incomes were more likely to meet the MDGs than the least developing countries. While perhaps not unexpected, this result is nevertheless disappointing. Countries whose peoples were most in need of assistance were, in fact, the ones least likely to receive it.

Whether the MDGs represent a continued dominance of the North over the South or the MDGs present a unique set of measurable goals and targets that will inspire countries to make faster progress toward development, this type of target has become a permanent feature of the global agenda in general and sustainable development. Hulme's [19] opinion that the MDGs will be ongoing and will eventually lead to the creation of other similar goal programs appears to be correct.

However, it is evident with the passage of time that the MDGs did not fully entwine the concepts of environment and development together. Given the rather diffuse origins of the MDGs, this is perhaps unsurprising. Perhaps, more importantly, the differences between a top-down approach to development rather than a bottom-up or hybrid approach to development should be examined. From the onset, the MDGs represented the idea that if donors focused enough time, energy, and resources on a specific subset of problems, developing countries could take in the time, energy, and resources to achieve a very visible set of concrete results. This is undoubtedly too simple a model.

11.4   Global Governance and the MDGs

One of the premises of this chapter is that the creation of the MDGs represents a radical departure from state-centric politics within the UN system to a model that also recognizes the individual as intrinsically valuable. While few today would question the strength of the moral argument represented by the MDGs, the departure from state-centric politics may not be as readily accepted, especially to proponents of neorealist thinking.

Thus, contemporary scholarship created a new conceptualization to describe this phenomenon called global governance. Importantly, global governance theory did not come about because of the MDGs, but rather the MDGs illustrate the phenomenon described by global governance. Finkelstein [14] aproclaims rgued for a broadening of the entire field of international relations by pointing out that predictable patterns of behavior emerge when entities and individuals adopt and follow norms and principles in the absence of a dominant enforcement scheme.

Global governance as a theory within international relations scholarship began in the late 1990s with the realization that non-state actors could alter state behavior [28]. Revolutions in both information technology and communication technologies fundamentally altered the workings of the international state system. The hierarchical structure where domestic actors sought control of their federal government who negotiated on their behalf rescaled horizontally as multinational corporations, international organizations, and NGOs realized the potential to bypass the state and communicate directly with their counterparts and vertically as global–local connections increased [2]. These communications ultimately led to decoupling norm selection and selection of preferred outcomes from traditional measures of power such as military might or economic wealth. Conceptualizations of power broadened to include technical knowledge, establishing discourse, and the role of shared norms [16, 24, 29].

In more practical terms, the recognition of global governance meant that states are no longer the only actor who matters [6, 7]. This revolutionary idea opened the door for proponents of new norms and standards of behavior such as the UN to build new partnerships with business and industry, and with prominent NGOs [30]. Consequently, a new UN constituency emerged, one based on embracing the idea that the transnational movement of goods and services came with a requirement to be perceived as a "good" global citizen, even if many corporations fall short of the ideal behaviors demanded of them [10].

This rescaling of international environmental affairs benefitted virtually all the major international actors. Developing countries received an immediate influx of financial aid, capacity building that created new infrastructure, and improved the quality of life for some of its citizens. International organizations such as the UN reestablished their importance in a global society that had begun to question whether funding the UN made sense in a world where the prospect of a permanent peace seemed closer than any point since its founding in 1945. Non-state actors received increased access to decision-makers that increased the possibility of their influencing outcomes. Whether the MDGs achieved their goal in spurring new levels of ODA or not, states, in general, and developing countries, in particular deemed the goal setting as meritorious. Consequently, the idea of promoting a global good via goal setting is, in all likelihood, here to stay.

References

1. Amin S (2006) The millennium development goals: a critique from the south. Mon Rev 57(10):1–15
2. Andonova LB, Mitchell RB (2010) The rescaling of global environmental politics. Annu Rev Environ Resour 35:255–282
3. Annan KA (2000) We the peoples: the role of the United Nations in the 21st century. United Nations, New York
4. Annan KA (2002) Toward a sustainable future. Environ Sci Policy Sustain Dev 44(7):10–15
5. Barnett M, Finnemore M (2012) Rules for the world. Cornell University Press, Ithaca
6. Betsill MM (2000) Greens in the greenhouse: environmental NGOs, norms and the politics of global climate change. University of Colorado at Boulder
7. Betsill MM, Corell E (2008) NGO diplomacy: the influence of nongovernmental organizations in international environmental negotiations. MIT Press, Cambridge

8. Biermann F et al (2017) Global governance by goal-setting: the novel approach of the UN sustainable development goals. Curr Opin Environ 26–27:26–31

9. Borowy I (2015) Negotiating international development: the making of the millennium development goals. Reg Cohes 5(3):18–43

10. De Bettignies HC, Lépineux F (2009) Can multinational corporations afford to ignore the global common good? Bus Soc Rev 114(2):153–182

11. Devarajan S et al (2002) Goals for development: history, prospects, and costs. In: Policy research working paper No. 2819 World Bank, Washington, DC

12. Dwyer-Lindgren L et al (2019) Mapping HIV prevalence in sub-Saharan Africa between 2000 and 2017. Nature 570:189–193

13. Easterly W (2009) How the millennium development goals are unfair to Africa. World Dev 37(1):26–35

14. Finkelstein LS (1995) What is global governance? Glob gov 1(3):367–372

15. Fukuda-Parr S (2010) Reducing inequality—The missing MDG: a content review of PRSPs and bilateral donor policy statements. Ids Bull-I Dev Stud 41(1):26–35

16. Haas PM (1990) Saving the Mediterranean: the politics of international environmental cooperation. Columbia University Press, New York

17. Haines A, Cassels A (2004) Can the millennium development goals be attained? Brit Med J 329:394–397

18. Hill PS et al (2010) Conflict in least-developed countries: challenging the millennium development goals. B World Health Organ 88:562

19. Hulme D (2009) The millennium development goals (MDGs): a short history of the world's biggest promise. BWPI Working Paper No. 100

20. Hurd I (2020) International organizations: politics, law, practice. Cambridge University Press, Cambridge, UK

21. Jones R (2013) 'Too many cooks in the kitchen,' warns MDG co-architect. In: Devex. https://www.devex.com/en/news/too-many-cooks-in-the-kitchen-warns-mdg-co-archit ect/80799. Accessed 27 Jan 2022

22. Kanie N, Biermann F (2017) Governing through goals: sustainable development goals as governance innovation. MIT Press, Cambridge

23. Karns MP et al (2015) International organizations: the politics and processes of global governance. Lynne Rienner, Boulder

24. Keck ME, Sikkink K (1998) Activists beyond borders: advocacy networks in international politics. Cornell University Press, Ithaca

25. Keita L (2020) Eurocentrism and the contemporary social sciences. Afr Dev 45(2):17–38

26. International Monetary Fund (2021) Report for selected countries and subjects. October 2020 Accessed at https://www.imf.org/en/Publications/WEO/weo-database/2020/October/ weo-re-port?c=111,&s=NGDP_RPCH,NGDPD,PPPGDP,NGDPDPC,PPPPC,PCPIPCH,& sy=2018&ey=2025&ssm=0&scsm=1&scc=0&ssd=1&ssc=0&sic=0&sort=oconut&ds=.& br=1. Accessed 17 Feb 2021

27. Langford M (2010) A poverty of rights: six ways to fix the MDGs. Ids Bull-I Dev Stud 41(1):83–91

28. Lipschutz R (1996) Global civil society and global environmental governance: the politics of nature from place to planet. SUNY Press, Albany

29. Litfin K (1994) Ozone discourses: science and politics in global environmental cooperation. Columbia University Press, New York

30. Martens K (2005) NGOs and the United Nations: institutionalization, professionalization and adaptation. Palgrave Macmillan, London

31. Mathiason J (2007) Invisible governance: international secretariats in global politics. Kumarian, Bloomfield

32. McArthur JW (2014) The origins of the millennium development goals. SAIS Rev Int Aff 34(2):5–24

33. OECD (1996) Shaping the 21st century: the contribution of development co-operation. OECD, Paris

34. OECD (2021) Our global reach. https://www.oecd.org/about/members-and-partners/. Accessed 17 Feb 2021
35. Poku N, Whitman J (2011) The millennium development goals: challenges, prospects and opportunities. Third World Q 32(1):3–8
36. Rapkin DP et al (1997) Institutional adjustment to changed power distributions: Japan and the United States in the IMF. Glob Gov 3(2):171–195
37. Richard F et al (2011) Sub-Saharan Africa and the health MDGs: the need to move beyond the 'quick impact' model. Reprod Health Matter 19(38):42–55
38. Rosenbaum B (2015) Making the millennium development goals (MDGs) sustainable: the transition from MDGs to SDGs. Harv Int Rev 37(1):62–64
39. Sachs JD (2012) From millennium development goals to sustainable development goals. Lancet 379(9832):2206–2211
40. Saith A (2006) From universal values to millennium development goals: lost in translation. Dev Change 37(6):1167–1199
41. Secretariat of the Basel Convention (2021) Basel convention milestones. http://www.basel.int/TheConvention/Overview/Milestones/tabid/2270/Default.aspx#:~:text=The%20Basel%20Convention%20plays%20a,us%20to%20achieving%20the%20MDGs. Accessed 22 Feb 2021
42. Stubbs P (2003) International non-state actors and social development policy. Glob Soc Policy 3(3):319–348
43. UNDP (2021) Millennium development goals. https://www.undp.org/content/undp/en/home/sdgoverview/mdg_goals.html. Accessed 21 Feb 2021
44. UN DESA (2015) The millennium development goals report 2015. United Nations, New York
45. Waage J et al (2010) The millennium development goals: a cross-sectoral analysis and principles for goal setting after 2015. Lancet 376(9745):991–1023
46. Weaver C (2008) Hypocrisy trap: the world bank and the poverty of reform. Princeton University Press, Princeton
47. World Bank Group (2021a) Understanding poverty overview. https://www.worldbank.org/en/topic/poverty/overview. Accessed 17 Feb 2021
48. World Bank Group (2021b) Millennium development goals. goal 1. http://www5.worldbank.org/mdgs/poverty_hunger.html. Accessed Feb 2021

Chapter 12
The WSSD

Abstract Occurring in the aftermath of the 9/11 attack on the United States, the WSSD ended the mega-conference trajectory within international environmental diplomacy. Intended to focus on implementation of Agenda 21, the WSSD did not attempt to match the ambitions of both the UNCHE and UNCED. More disappointingly, states could not achieve consensus on implementation. Thus, WSSD turned to the Type II partnerships to salvage the meeting. Scholarship following the WSSD pointed to numerous reasons for the failure, including the least common denominator position necessary to achieve consensus (Gutman in Environ Sci Policy Sustain Dev 45:20–28 [11]), conference fatigue (Wapner in Global Environ Polit 3:1–10 [23]), economic liberalization stronger norm than sustainable development (La Viña et al. in SAIS Rev 23:53–70 [14]; Wapner in Global Environ Polit 3:1–10 [23]), difficult geopolitical climate (Gutman in Environ Sci Policy Sustain Dev 45:20–28 [11]; Wapner in Global Environ Polit 3:1–10 [23]), and lack of political will (Gutman in Environ Sci Policy Sustain Dev 45:20–28 [11]; Mestrum in Environ Dev Sustain 5:41–61 [15]; Von Frantzius in Env Polit 13:467–473 [22]). To a certain extent, any or all of these could be sufficient for any meeting to fall short of expectations. The WSSD, unfortunately, exhibited all these characteristics.

Keywords World Summit on Sustainable Development · Mega-conference · Johannesburg plan of action · Type II partnerships · Diplomatic failure

In the aftermath of the attacks on the World Trade Center in New York, United States, global diplomats gathered in Johannesburg, South Africa, to celebrate the successes of international cooperation and the improvement in environmental quality at the WSSD. The primary focus of the meeting was to move toward the sustainable development paradigm, including determining what actions might be needed to ensure that these systemic changes further improved the quality of life for all peoples.

That the global North turned toward embracing developmental concerns at all should have, by all rights, been considered a success given the historical emphasis on traditional environmental concerns such as clean air, clean water, and healthy soil. This endorsement, however, brought about new problems, that sustainable development could be manipulated to represent the current economic status quo. Ever-increasing amounts of economic trade globalization brought with it increasing social inequalities and its unwelcome environmental consequences that highlighted Southern skepticism.

While the UN General Assembly undoubtedly wanted Johannesburg to be an action-oriented conference that vaulted sustainable development from a soft law principle to a regime, the current geopolitical status gave little reason to believe that any meaningful results would be achieved. The North–South financial gap had grown wider, environmental pollution continued at unprecedented rates, and foreign aid decreased in real terms and as a percentage of Gross Domestic Product [9].

The United States, in particular, had undergone a significant shift in foreign policy with the change of its presidency from President Clinton to President Bush in January 2001. President Bush differed from President Clinton on many domestic and foreign policies. President Bush believed in the unrestricted use of American military and economic power, including the willingness to forge ahead without its allies. The United States was unlikely to support any agreement that might come out of the WSSD process. Further, the world's one remaining superpower was fighting a war with Afghanistan after suffering from a devastating attack at home in its commercial center, New York City, and its political capital, Washington DC. The September 11th attacks on the World Trade Center and Pentagon lowered American interest in attempting to change the socioeconomic system it was currently defending through military action.

Expectations for the WSSD hit an all-time high. Proponents of sustainable development saw, and continue to see, every conference as a potential breakthrough moment that will create a tipping point for the adaptation of sustainability. As a consequence, every meeting that fails to meet this admittedly high standard is harshly critiqued; the WSSD is no different. Thus, observers familiar with the rhythms of the mega-conferences initially hesitated to declare the WSSD a failure. It did not create the tipping point necessary to propel sustainability dramatically forward. It also was not a failure. The UN system made an incremental move toward sustainability due to this conference. This movement, however, did not occur because of state action. Non-state actors and the UN showed an increased interest in launching partnerships on a variety of issues within global environmental affairs [1].

Whether as a result of the WSSD directly, the early years of the twenty-first century did not launch a millennium of global cooperation between states. Indeed, much of the next decade saw a formal diplomatic retreat on environmental affairs rather than a steady move forward. Non-state actor activity increasingly gained recognition within global society, one capable of delivering on the UN's preferred policy actions. Thus, UN recognition of non-state actors' willing support and participation created a significant structural addition to formal diplomatic powers.

This brief chapter contains two sections. Section one opens this chapter with a brief overview of the events leading to the WSSD and its lackluster conclusion. Section two presents an overview of the failure to reenergize international environmental diplomacy after this meeting.

12.1 Johannesburg

True to UN format during the Stockholm-Rio-Johannesburg mega-conference trajectory, the WSSD conference consisted of four PrepComs and a two-week conference. The initial organization of the meeting also comprised a large number of other regional meetings and a series of informal consultations away from the PrepCom sessions. Unlike its predecessors, Johannesburg suffered from an overly complex negotiating agenda as it also encompassed significant portions of the agenda from past conferences. These conferences included the Millennium Development Summit, the MDGs, the Doha Declaration from a WTO meeting that affirmed the role of the current liberal economic system in creating economic growth, and the Monterrey Consensus, where the United States and the EU pledged $30 billion for sustainable development [13].

In a dramatic departure from the preparatory processes for Stockholm and Rio, the run-up to Johannesburg was significantly less intensive than that which occurred for its predecessors. This early development signaled that countries' ambitions for the conference were less than previous conferences. This relative lack of intensity was signified by the shortened time scheduled for the preparatory meetings and the double-duty where the tenth meeting of the CSD also served as PrepCom I. First, the number of formal negotiating days for Johannesburg was less than half of the number allocated for Rio and roughly three-fourths the number in Stockholm. This time frame limited the opportunity to find common ground before the conference. Second, the dual nature of the CSD serving as the first PrepCom for WSSD also signified the lack of political importance member states attached to Johannesburg. The dual use of the CSD as the PrepCom did not limit the number of countries as states elected to use open ended rules of participation [21].

PrepCom I thus served not only to organize countries for the remainder of the negotiations it also determined the main themes for the WSSD itself. Environmental themes such as atmosphere, oceans, water, waste, and energy were carried forward from past conferences. Similarly, the MDGs' elements of poverty, health, and education also remained on the proverbial negotiating table. Additionally, diplomats deemed items such as sustainable tourism, consumption patterns, and the role of major groups within civil society, such as businesses, Indigenous Peoples, NGOs, women, and youth among others, ripe for discussion as well [21].

Preparations for Johannesburg continued through the PrepCom process and associated regional meetings. Although, by this time, the Johannesburg summit had one year to negotiate in advance of the main conference, while both the Stockholm and the Rio conferences had two years. Advice about the conference also came forth from the UN Secretary-General and the expected position papers and press releases from civil society. The PrepCom process began in the spring of 2001, albeit with an unusual aura of cynicism. States elected Emil Salim (Indonesia) as PrepCom chair and moved to begin consideration of the 24 documents provided by the UN Secretariat for this meeting. The opening day saw participants questioning the necessity of the meeting [7] and recommended that the conference minimize its ecological footprint. NGOs were concerned about the logistics and organizational arrangements in Johannesburg. After the physical separation imposed in Rio, NGO participants were particularly anxious for close physical proximity to the formal conference.

Despite the questions surrounding the necessity of WSSD, the organizational session eventually adopted an overall positive atmosphere. PrepCom I quickly voted in favor of agreements on the structure of the meetings, including dates and locations [7]. However, the delegates did not comply with Chair Salim's request to discuss the materials distributed by the UN Secretariat. This waste of valuable meeting time further reduced the likelihood that states would come to any type of meaningful agreement by the end of the WSSD negotiating process. Compared to the other PrepCom meetings, WSSD ended its first session on a more positive note than its opening plenary but concluded the least amount of business. Delegates had hardly discussed the North–South environment/development gap, much less reconciled the divergent opinions on an exceptionally broad agenda. Additionally, participants speculated about the effects of the US refusal to sign the Kyoto Protocol that contained binding targets for reducing greenhouse gases [5].

PrepCom II met from January 28–February 8, 2002, in New York. During this meeting, PrepComII began consideration of the documentation generated advanced from the first session. Additionally, another eighteen background papers had been forwarded for PrepCom's consideration. PrepCom II spent the first week generating dialogue and topics to be included in the debate. Chair Salim consolidated the topics into a Chairman's paper that delegates agreed to use this text as the basis for negotiations at PrepCom III. The issues included in the Chairman's paper were: poverty eradication, changing unsustainable patterns of consumption and production, protecting and managing the natural resource base of economic and social development, sustainable development in a globalizing world, health and sustainable development, sustainable development of small island developing states, sustainable development initiatives for Africa, means of implementation and strengthening governance for sustainable development at the national, regional, and international levels [21].

Chair Salim also emphasized the outcomes for this conference as poverty eradication, consumption and production, and resource conservation for development [18]. The issues included in the consolidated chair paper reflect the highly controversial norms of equality of wealth and industrialization. From the start of the WSSD negotiations, the wealthier states were not likely to accede to the demands made by the G77 and China for increased developmental aid, even if these states couched these requests for further assistance in the less controversial agenda items of environmental health and protection. In addition to the consolidated chair paper, NGOs tabled text calling for the negotiation of standards for corporate responsibility. This initiative proved timely as the Enron and WorldCom scandals were headline news. The text about corporate responsibility survived the remainder of the PrepCom negotiations to Johannesburg, where conferees tabled it along with the majority of the other action-oriented items.

PrepCom III met from March 25–April 5, 2002 in New York. The participants' order of business was to discuss the Chair's Paper drafted during the previous meeting. The goal was to turn this document into a consolidated negotiation text assessing institutional frameworks for sustainable development and strengthening these organizations. PrepCom III was not successful in this attempt as logistical and organizational problems consistently plagued the meeting. Indeed, a UN common services decision at the end of the first week to keep the rooms open after 6:00 pm was included in the ENB, the "unofficial" daily newspaper for the Conferences [8]. The meeting's poor organization, lack of UN resources for logistical support, and poor preparation, particularly from the G-77/China, resulted in a notable lack of progress and diminished the likelihood for a meaningful WSSD [8].

In addition to the Chairman's Paper, two other topics received significant attention; a discussion paper entitled "Sustainable Development Governance at the International, Regional and Global Levels" and Type 2 partnerships between governments and non-state actors to strengthen sustainable development. Type 2 partnerships are voluntary collaborations between governments and non-state actors that focus on environmental health and protection, while Type 1 partnerships involve state aid directly to another state. By the end of PrepCom III, delegates promoted Type 2 partnerships into a major outcome of the WSSD, as the likelihood of achieving any other kind of agreement declined rapidly. Secondary importance to the Type 2 partnerships stemmed from the mindsets of many conferees, namely, that commitment to sustainable development had become synonymous with development aid, measured in terms of monetary contributions. Johannesburg was not going to be a meeting to develop or debate new norms or the expansion of sustainable development.

PrepCom IV met from May 27–7 June 7 in Bali, Indonesia. Two days of informal negotiations preceded the conference to discuss the Revised Chair's Paper. Significant progress was evident as conferees gradually transformed the Revised Chair's Paper into a Draft Plan of Implementation [21]. However, negotiations at Bali did not go smoothly. Text thought to be finalized, could be renegotiated, states complained about being excluded from informal consultations, and support staff did not appear to be knowledgeable when asked for assistance [2]. The delegates discussed elements for a political declaration but could not agree upon a draft text. As a result, Chair Salim agreed to prepare a negotiating text for Johannesburg.

The lack of finalized negotiating text for Johannesburg, while not unexpected given the lack of progress in earlier meetings, nevertheless signified the weakness of the WSSD as well as provided an indication of the vast difference of opinion regarding the future of sustainable development [2]. More importantly, perhaps, than the current agenda items were the items that had been moved off the table, such as establishing a world environmental organization and utilizing the conference itself as a deadline for entry into force of several multilateral environmental agreements. These treaties included the Kyoto Protocol dealing with climate change, the Cartagena Protocol, the Rotterdam Convention, and the Stockholm POPs treaty. This lack of preparation turned into low expectations for the Summit in Johannesburg and led to a discussion of an NGO boycott, particularly among environmental organizations [11].

The promotion of the Type 2 partnership to a major conference outcome not only demonstrated a failure to make significant headway on Agenda 21 in the past ten years but also indicated a shift of emphasis from state action to voluntary action. It simultaneously elevated the importance of NGOs and other non-state actors within the sustainable development paradigm, as they became an explicit target of the negotiations. Norris [17] performed a preliminary analysis on the Type 2 partnerships negotiated as part of the WSSD process. Her report shows a surprising lack of participation from certain actors, most notably the low presence of businesses and the absence of China and India. To a certain extent, the partnerships were designed to hide the fact that the Johannesburg conference was largely unsuccessful.

Wapner [23] agrees that the Type 2 Partnership became the primary success for the WSSD. These partnerships were born of a necessity for the UN to deliver some kind of financial assistance to the underdeveloped South. However, it is generally not clear whether the funds from the projects are new and additional. Partially, as a result of this, the partnerships themselves came under increased criticism. By the end of the meeting, corporations, in particular, had shied away from publicizing new commitments.

Despite these problems, NGOs benefited from improved access to the PrepCom sessions. Numerous NGOs spoke during meetings and made recommendations for text. Major groups spoke during the second and fourth meetings, although none of the speeches shared a common theme or issue. The WSSD Secretariat made a special effort to involve the major groups at Johannesburg by including them in roundtable discussions, in addition to the "normal" speeches and sidebar events coinciding with a major UN conference. That is not to say the NGOs had free reign in the lead-up to Johannesburg. States initially delayed accreditation for the Tibet Justice Center after China objected to its participation because the Tibetan group supported separation from China [2]. Tibet Justice Center lost its argument for accreditation at PrepCom IV, and this decision served as a reminder that NGOs must walk a fine line between freedom of speech and offending UN member states.

Despite these pre-meeting setbacks, the WSSD held on August 26–September 4, 2002, in Johannesburg, was a significant global summit. However, its achievements are not nearly as notable as Stockholm or Rio. The meeting produced two important documents—the Declaration on Sustainable Development and the Plan of Implementation. Both documents represented major setbacks for the sustainable development agenda. Gutman [11] criticizes both documents for failing to move beyond pre-existing international agreements.

Overall, the Summit kept sustainable development alive on the international agenda by repeating reassuring words of governmental support. However, governments attending the conference were unwilling to make meaningful long-term commitments to this international ideal. The WSSD documents represented a "business-as-usual" attitude that prioritized traditional conceptualizations of the global economy with an emphasis on trade liberalization and globalization. Ironically, one major success for the environmental movement occurred with the failure to designate the WTO as the appropriate forum for reconciling conflicts between environmental treaties and trade agreements. If this move had been finalized, it could have moved environmental affairs under international trade. This move also would have limited the future role of NGOs in global politics as the UN grants NGOs more privileges than the WTO. In a more cynical vein, the fact that this item was on the agenda could, in and of itself, be viewed as a significant setback for pro-environment conferees.

The South African Delegation circulated draft text for the Political Declaration, based upon the "elements" agreed to during PrepCom IV at Bali. Time constraints prevented significant negotiation, and the Chair's text was largely left intact. The document provides a brief historical analysis from Stockholm to Johannesburg before outlining challenges to sustainable development. The document concludes by reaffirming its commitment to sustainable development and multilateral solutions, including the UN. The fourth section of the Johannesburg Declaration, *Our Commitment to Sustainable Development*, captures the essence of the conference with its focus on the discourse of sustainable development and the process of embedding it within the international system [25]. Because the Declaration summarizes political thought about sustainable development at the time of the WSSD, its usefulness to inspire future negotiations is questionable. It largely fails to go beyond broad generalizations.

Negotiations on the Plan of Implementation began at the last minute in the informal negotiations held immediately before the opening of the WSSD at Johannesburg. Wapner [23] stated that Johannesburg had the unenviable task of addressing implementation, an issue that other conferences could not resolve. The attempt to address the topic came about as the Plan of Implementation, a wide-ranging document that consists of eleven chapters. The preamble provided the historical context and references to the Rio Declaration, Agenda 21, the MDGs, and the Monterrey Declaration that dealt with partnerships. The Plan of Implementation also highlighted key themes for this conference such as poverty eradication, common but differentiated responsibilities, good governance, sustainable consumption and production, partnerships with civil society, globalization, and the need for "peace, security, stability, and respect for human rights and fundamental freedoms, including the right to development" [25: 9]. The remaining chapters addressed the main issue areas, including protection and management of natural resources, health, small island developing states, Africa, other regional initiatives, implementation, and institutional framework.

Once again, issue areas at Johannesburg split into the North's preference for technical environmental protection and the South's preference for industrial development. Chap. 5, dealing with globalization, in the Johannesburg Plan of Implementation was predictably contentious, as it dealt with the interlinkages between trade, finance, and environment. While not on the agenda per se, a great deal of handwringing also occurred because of the United States refusal to ratify the Kyoto Protocol and what this might mean for the future of the climate change regime in the absence of the world's superpower and one of its leading emitters of greenhouse gases.

In addition to the main conference at Johannesburg, three other venues garnered significant attention—the Water Dome, the Ubuntu village, and the Global Peoples Forum in Nasrec. Contemporary commentaries on the Johannesburg Summit credit all three of these conferences with conducting a more eloquent discussion on sustainable development than the official venue at Sandton, an upscale Johannesburg neighborhood [13]. More meaningfully, both NGOs and businesses used this event as an opportunity to strengthen their relationship with the UN. While NGOs did not attend the Summit in the record numbers as its predecessor conference in Rio. WSSD nevertheless garnered the attention and participation of more than 8000 observers from 1204 accredited NGOs. Additionally, governments, NGOs, and private businesses announced the formation of 220 partnerships during the conference that totaled around $235 million to create practical demonstrations of sustainable development [16]. Gutman [11] disputes these numbers by highlighting that the partnerships' database included duplicates and estimated that the actual number of partnerships could be as low as 110.

Scholars typically disagree about the importance of negotiating position shifts and their significance during the Rio Conference. Wapner [23] suggested that the South shifted its position to embrace environmental issue areas in addition to development areas. He also pointed out that the United States shifted its position to champion economic globalization at the expense of all other social norms.

Egelston [5] believed that the most notable difference between Rio and Johannesburg involved the addition of social justice to environmental concerns. She pointed out that the discourse on sustainability is more likely to include strong statements for the social pillar in addition to the environmental pillar. For example, the Johannesburg Declaration mentions "the need for human dignity for all," "the indignity and indecency occasioned by poverty," "the need to produce a practical and visible plan to bring about poverty eradication and human development," before mentioning natural resources, biodiversity loss, desertification, and climate change [25: 1]. This language reinforces the shift in the meaning of sustainable development away from the North's preference for environmental protection and toward the South's preference for industrial development.

While earlier conferences focused on broad goals such as improving the quality of life through eco-development, the WSSD emphasized specific, technical goals including—access to potable water, improvement in sanitation systems, and housing that the developing countries frequently lack. While these easily quantifiable goals are laudable, the likelihood of developed countries increasing funding to provide these necessities outside of their borders is perhaps unrealistic given the lack of equity of wealth distribution within the North. Consequently, WSSD failed even to produce the inspiring rhetoric of earlier conferences.

12.2 Why Do Conferences Fail?

WSSD ushered in an unwelcome era within international environmental diplomacy. Progress on further elaborations of existing protocols and new soft law principles slowed dramatically. Bigg [1] explained that determining the success of the WSSD depended, in part, on the benchmark utilized. Scholars that focused on formal outcomes were likely disappointed [6, 19, 24]. Others focused on the bevy of activity, especially involving non-state actors, might find the WSSD more motivating and successful [10]. Others provided a more nuanced views as they concluded that among the non-state actors, business and industry groups viewed the WSSD more positively than environmental NGOs [3, 12].

If one was looking for the WSSD to create a world environmental organization that could successfully compliment (or compete with) the WTO, then WSSD was an abject failure. In the past, diplomatic actions have been taken to avoid declaring a conference a failure. One example of this included the suspension of COP-6 (The Hague) to avoid admitting that countries failed to agree on carbon cap and trade agreements under the Kyoto Protocol, as detailed in Chap. 8.

In extreme circumstances, diplomatic failures have led to the continuation of wars. In recent times, conference failures refer to the fact that the conference did not create a pre-determined goal. Goals could include producing a new hard law treaty, complete with targets and timetables, or advancing a soft law principle. These goals would be normative shifts that are at the heart of the international environmental agenda.

The academic literature is replete with reasons for this failure at WSSD, including the least common denominator position necessary to achieve consensus [11], conference fatigue [23], economic liberalization stronger norm than sustainable development [14, 23], a difficult geopolitical climate [11, 23] and lack of political will [11, 15, 22]. To a certain extent, any or all of these could be sufficient for any meeting to fall short of expectations. The WSSD, unfortunately, exhibited all these characteristics.

While the WSSD event did not suffer the international embarrassment associated with COP 6, the shift in the PrepComs to the least common denominator consensus position did signal the end of the easy victories for sustainable development. Recalling that international environmental negotiations tend to be dominated by three negotiating blocks, the least common denominator consensus position meant that each negotiating block could not agree on a path forward. To maintain unity within the block, each group weakened its negotiating stance. This, in turn, meant that advancing new principles, timetables, or targets did not have a state willing to champion the cause.

Another explanation for the diplomatic failure at Johannesburg could be that conference fatigue set in as a result of the ambitious international negotiating agenda that saw a minimum of one major conference a year [11, 23]. While on the surface, a two-week negotiating session once a year might not seem strenuous, this superficial reasoning belies the significant work that goes into preparing and negotiating a conference. While diplomats shortened the PrepComs for WSSD, other meetings continued on a "normal" schedule. In addition to the travel and the in-person negotiating sessions, diplomats also hold informal consultations with other countries and prepare numerous talking points and position papers to brief other members of the negotiating team.

Underlying tensions between the current economic system promote economic globalization and downplay sustainable development. Wapner [23] pointed out that the international trade regime is significantly stronger than an environmental protection regime during this time frame. One way this manifests itself is by weakening fundamental environmental norms. For example, the precautionary principle established previously at Rio did not reappear in the same form at Johannesburg [14]. This failure potentially represents a major setback for environmental protectionism. The precautionary principle effectively shifts the burden of proof onto manufacturers to prove that their products are safe for use. When this principle is removed from environmental affairs, it shifts from a protective stance to a reactionary stance, meaning that manufacturers may introduce new items at will with the understanding that those goods and services that damage the environment will cease, if necessary, in the future. Other non-agenda items at Johannesburg included the role of advertising in promoting unsustainable consumption, military spending, and adverse environmental impacts.

In light of the scaling back of both issue areas and of commitments under the issue areas discussed, Southern states looking for the WSSD to provide additional funding for development quickly found those hopes dashed. Gutman [11] points out that additional financial assistance from states was slow to emerge and that poverty reduction did not receive any new pledges of aid.

Wapner [23] laments the change of heart of the United States from environmental champion to environmental laggard. Once at the forefront of the environmental movement, the United States at WSSD proclaimed that economic globalization could solve many sustainable development problems. Nor was the United States alone in stepping back from environmental issues, in that the EU did not immediately assume a leadership role either. In other words, two of the three major diplomatic negotiating blocks did not demonstrate a strong political will to advance sustainable development during the run-up to the Summit nor during the Johannesburg meeting.

Mestrum [15] opined that this lack of political will could be seen when evaluating the poverty reduction policies recommended in the Johannesburg Plan of Implementation. She noted that the policy recommendations on this topic represent a clear step-back from Agenda 21, negotiated at Rio. Nor is this the only issue area that failed to receive the same level of political support as previous UN documents.

In addition to the formal negotiating sessions, the follow-up to the conference also determines the success (or failure) of diplomatic negotiations. Consequently, no assessment of a mega-conference could be deemed complete without evaluating post-conference implementation. Implementation could be construed as consisting of the actions taken by states in carrying out their self-imposed tasks. After all, the realist model of international relations theorizes that states are the primary actor at the international level. Agreements made between states are then left to the state to enact at the domestic level. Given that Johannesburg created very few new targets or timetables for states to enact domestically, tracing any concrete environmental impact from the WSSD is impossible.

However, the liberalist viewpoint sees individuals and the groups they create as actors as well. In this sense, NGOs, especially those that possess technical knowledge on environment and development, become primary actors in addition to states. Under this model, non-state actors may be asked to undertake post-conference implementation tasks. Consequently, any review of post-conference implementation should include these non-state actors as well. As with states above, tracing a concrete environmental impact from a non-state actor to an environmental benefit would be a herculean task as well. However, the creation of the Type 2 partnerships opens room for speculation that the UN's willingness to acknowledge the works of both private industry and civil society may promote these groups during this time frame [4].

Whether a UN conference stands as a success or a failure, it is nevertheless crucial to understand past negotiations. International environmental diplomacy moves slowly, more often than not. Diplomatic forays often involve incremental movement forward with a variety of actor types representing diverse interests. While conferences may not meet observer or state expectations on delivering specific outcomes, learning from failures impacts the future, albeit in a different way than learning from successes.

References

1. Bigg T (2002) The World Summit on Sustainable Development: an assessment. international institute for environment and development. http://www.jstor.org/stable/resrep16732. Accessed 29 Jan 2022
2. Buenker MA (2002) Progress at Bali but not enough for Johannesburg. Environ Policy Law 32(3–4):140–151
3. Burg J (2003) The World Summit on Sustainable Development: empty talk or call to action? J Environ Dev 12(1):111–120
4. Carr DL, Norman ES (2008) Global civil society? The Johannesburg World Summit on Sustainable Development. Geoforum 39(1):358–437
5. Egelston AE (2007) Sustainable development: a history. ProQuest
6. Eichenberg T, Shapson M (2004) The promise of Johannesburg: fisheries and the World Summit on Sustainable Development. Gold Gate Law Rev 34(3):587–644
7. ENB (2001) Prepcom-1 highlights Monday, 30 Apr 2001. 22(1):1–2
8. ENB (2002) Summary of the third session of the preparatory committee for the World Summit on Sustainable Development: 25 March–5 April 2002. 22(29):1–13
9. Elnur I (2003) 11 September and the widening north-south gap: root causes of terrorism in the global order. ASQ 25(1/2):57–70
10. Gray KR (2003) World Summit on Sustainable Development: accomplishments and new directions? Int Comp Law Q 52(1):256–268
11. Gutman P (2003) What did WSSD accomplish? An NGO perspective. Environ Sci Policy Sustain Dev 45(2):20–28
12. Hamann R et al (2003) Responsibility versus accountability? interpreting the World Summit on Sustainable Development for a synthesis model of corporate citizenship. J Corp Citizenship 9:32–48
13. Hans L, Nath B (2005) The Johannesburg conference. In: Hans L, Nath B (eds) The World Summit on Sustainable Development: the Johannesburg conference. Springer, Dordrecht, pp 1–34
14. La Viña AG, Hoff G, DeRose AM (2003) The outcomes of Johannesburg: assessing the World Summit on Sustainable Development. SAIS Rev 23(1):53–70
15. Mestrum F (2003) Poverty reduction and sustainable development. Environ Dev Sustain 5(1):41–61
16. Middleton N, O'Keefe P (2003) Rio plus ten politics, poverty and the environment. Pluto Press, London
17. Norris C (2005) Partnerships for sustainable development the role of type II agreements. In: Churie A et al (eds) Global challenges furthering the multilateral process for sustainable development. Greenleaf Publishing Sheffield, pp 210–230

18. Salim E (2002) A journey of hope statement by the chairman of the preparatory committee for WSSD, Mr Emil Salim on the final day of the second session of the committee New York 8 February 2002. 8 Feb 2002, New York

19. Scherr S, Gregg R (2006) Johannesburg and beyond: the 2002 World Summit on Sustainable Development and the rise of partnership. Georget Environ Law Rev 18(3):425–464

20. UN GA (2001). Report of the Commission on Sustainable Development acting as the preparatory committee for the World Summit on Sustainable Development organizational session (30 April–2 May 2001). A/56/19

21. UN GA (2002). Chairman's paper. A/CONF.199/PC/L.1

22. Von Frantzius I (2004) World Summit on Sustainable Development Johannesburg 2002: a critical analysis and assessment of the outcomes. Env Polit 13(2):467–473

23. Wapner P (2003) World Summit on Sustainable Development: toward a post-Jo'burg environmentalism. Global Environ Polit 3(1):1–10

24. Wilson M (2005) The new frontier in sustainable development: World Summit on Sustainable Development type II partnerships. Vic Univ Wellingt Law Rev 36(2):389–426

25. WSSD (2002) Report of the World Summit on Sustainable Development Johannesburg, South Africa, 26 Aug–4 Sept 2002. A/CONF.199/20

Chapter 13
Climate Change, Redux

Abstract The withdrawal of the United States jeopardized climate mitigation actions during the 2008–2012 first commitment period. It also heightened fears that climate catastrophes were closer than ever before. While the European Union moved into a leadership role in reducing greenhouse gas emissions, international diplomats moved forward with implementing the Kyoto Protocol while also looking for a pathway to reengage the United States. This futile effort further weakened climate diplomacy resulting in several notable failures including the Copenhagen Accord, and to a lesser extent, the Paris Agreement. To date, no legally binding climate mitigation protocol has been adopted by the international community of states. Meanwhile, academic scholarship more closely examined the processes by which regimes formed, noting that climate change presented unique opportunities to study the interlinkages between regimes. Scholars created a new concept, regime complexes, to describe this situation.

Keywords Climate change · Copenhagen Accord · Paris Agreement · Regime complex · Carbon trading · Climate mitigation

After the lackluster WSSD covered in Chap. 12, global support for additional environmental treaties notably declined. As the United States focused global politics on the War on Terror, states left the environment behind. The so-called "peace dividend" from the end of the Cold War never materialized as traditional international concerns of violence and regional instability resumed their usual location at the top of the hierarchy of global issues. Further, the War on Terror created environmental damage, including the destruction of the natural environment from bombs, rockets, and landmines, while destroyed buildings further exposed people to the harsh reality of living in a war zone [34]. United States soldiers were not exempt from facing environmental health hazards. Those sent overseas as part of this military action returned as victims of toxic air pollution from burning trash in open air pits [33].

While the War on Terror unquestionably added to the amount of global environmental destruction, this military engagement did not end diplomatic negotiations on environmental issues or end states' compliance with existing international environmental treaties. Rather, conversations continued, but the absence of the United States' environmental leadership meant that countries were further apart rather than closer

© The Author(s), under exclusive license to Springer Nature Switzerland AG 2022   201
A. Egelston, *Worth Saving*, AESS Interdisciplinary Environmental Studies
and Sciences Series, https://doi.org/10.1007/978-3-031-06990-1_13

together. This increased tension stemmed from multiple facets of limiting climate change. Perhaps none more important than the realization that any future climate change regime could allow other countries to lower costs of fossil fuels, effectively ending one of the United States' primary economic advantages. Worse yet, the terms and conditions of a future climate change regime meant that other competitor countries, noticeably China, would not be limited in their ability to burn fossil fuels, thus gaining an economic advantage over the United States.

While it is convenient to point the proverbial finger at the two Republican presidents of the United States, George W. Bush, and Donald Trump, as primarily responsible for the lack of progress on combatting climate change after 2001, responsibility for the lack of movement should be spread throughout the world. Countries in the Middle East never had a strong incentive to support an environmental treaty that would significantly curtail the use of their primary export—oil. Further, Russia, the United States, and China all believe that a climate change regime may potentially limit their domestic economies in the short term, if not the long term. Consequently, climate change politics focus as much on economics as on the protection of the global atmosphere as the transition to a low carbon economy could fundamentally alter the balance of power between states [15].

In contrast to the reluctance of states to take on binding reductions of carbon emission, the Alliance of Small Island States (AOSIS) and other low-lying countries, including a significant portion of Northwestern Europe, believe that their continued existence is eminently threatened by climate change. For example, 17% of the Netherlands' landmass is currently below sea level, and the country maintains an extensive flood control system to keep water away from usable land. Any additional sea level rise could overwhelm the current flood control system and result in a catastrophic change to the people currently living in this area.

This chapter continues the story of the climate change negotiations by reviewing the treaty contents, negotiating history, and scholarly theories after COP 6bis (the resumed COP 6 meeting) in Bonn, Germany. The United States withdrawal from the Kyoto Protocol in 2001 jeopardized climate mitigation actions during the 2008–2012 first commitment period. It also heightened fears that climate catastrophes were closer than ever before. The absence of the United States not only made it more likely that climate reductions would be more difficult to achieve, but it also led to concerns about future financing as the United States is a wealthy country that finances a significant amount of international aid and assistance, even if it fails to meet the 0.7% of gross national income ODA standard created by the international system in the 1970s, but never officially accepted by the United States.

The withdrawal also created a leadership void that was essentially filled by the EU. The EU included fifteen members throughout most, but not all of Western Europe, so this supranational authority wields significant political power. Further, the EU's decisive domestic action on climate change increased its credibility at the international level. Thus, it is necessary to remember that various levels of analysis, the global, the supranational, the national, and the local, do not occur separately. They are interlinked.

13.1   COP 6bis and COP 7

As Chap. 8 ended, a stalemate between the United States and the EU at COP 6 resulted in the suspension of the negotiating session without an agreement. Thus, COP 6 occurred in two parts, the initial meeting at The Hague, Netherlands, and the resumption of the negotiating session, designated COP 6 bis, at Bonn, Germany. In the meantime, the United States confirmed the election of George W. Bush, Governor of Texas. The resulting change in power within the United States essentially meant that the EU gamble to resume negotiations with their preferred negotiating partner, Al Gore, in the hopes of negotiating stricter reductions in climate emissions, failed.

As discussed in Chap. 8, the Kyoto Protocol represented a compromise between the EU's preference to reduce carbon emissions through the cooperative use of policies and measures versus the United States' preference to rely on market mechanisms. After the withdrawal of the United States, the EU could have placed greater emphasis on the use of policies and measures. Instead, the EU and others chose instead to focus on the Kyoto market mechanisms. Consequently, international diplomats focused their efforts on operationalizing the Kyoto mechanisms The United States undeniably lost a major opportunity to craft the future of the carbon markets when COP 6bis resumed in Bonn, Germany, from July 16–27, 2001.

The Bonn Agreements emerged as the primary outcome of COP 6bis, a carefully crafted series of clarifications that added an additional layer of detail to the Kyoto Protocol by providing further information on four key topics such as the role of sinks, compliance, financing for developing countries, and the operationalization of the Kyoto Mechanisms [29, 36]. Vrolijk [41] believed that the Bonn Agreements occurred only because the United States withdrawal from the Kyoto Protocol had the ironic impact of encouraging other countries to renew their efforts to create a concrete climate change scheme.

The first key arrangement that emerged from the Bonn Agreement involved one of the major points of contention between the United States and other members of the Umbrella Group, and the EU; that is the stringency of the Kyoto commitments concerning the use of carbon sinks. In fact, Ott [29] believes that this issue ultimately caused negotiations at COP-6 in The Hauge, Netherlands, to unravel. Including carbon sinks in the carbon accounting cycle potentially lowers the number of reductions each existing source might need to take in the future, negating the need to find fewer intensive sources of using energy [8]. This situation, in turn, potentially weakens the environmental effectiveness of the Kyoto Protocol [29].

Carbon storage, or sinks, occurs when carbon is absorbed into the natural environment and stored for long periods of time. While carbon sinks may occur naturally in oceans, soils, and forests, one particularly noteworthy form of carbon sinks includes trees absorbing carbon during photosynthesis. Carbon sinks may also be manmade. Theoretically, corporations could create a carbon reservoir by injecting carbon far enough underground that it would be unable to reach the atmosphere in a process known as carbon sequestration. However, significant challenges with carbon sequestration remain unsolved, most notably the high energy costs associated with this process. Carbon sinks, therefore, could be created by planting new forests. Allowing states to create new carbon sinks meant that some states could avoid making reductions at industrial facilities, a highly undesirable outcome for many countries at risk due to rising sea levels. Storing carbon also creates risks of a large-scale carbon release because the carbon reservoir ceased functioning correctly. In the case of forests, a forest fire would be sufficient to trigger a significant release of carbon dioxide into the atmosphere. Additionally, carbon storage that occurs through many natural cycles continues to be poorly understood from a scientific perspective causing uncertainty in carbon emissions inventories, including both baseline calculations and future compliance calculations. Consequently, diplomats hesitated to create a compliance regime based upon the use of sinks. The final agreement stipulated that carbon sinks would not be included in the baseline but allowed limited use of sinks related to land-use, land-use change, and forestry in the Clean Development Mechanism.

Second, parties made significant forward progress regarding arrangements for monitoring compliance, including taking actions regarding instances of non-compliance. Legally binding agreements typically mean that harmed parties have the right to sue in a court of law to demand compliance. Legally binding agreements may also contain punishments such as fines to deter poor behavior. Typically, environmental treaties with strict terms for compliance are avoided by states as they do not wish to hand control over to a third party. Thus, the nature of the compliance agreement within the Bonn Agreement does not break new ground within international diplomacy.

States granted compliance oversight to a committee with two functions—facilitating compliance and enforcement. Both developed and developing countries may be subject to the facilitated compliance branch. Facilitating compliance techniques occur when the Compliance Committee helps states comply by providing technical assistance. The enforcement branch appears to be more focused on assuring that Annex I countries meet their reduction targets. Penalties for non-compliance include limiting countries from utilizing the Kyoto mechanisms and forcing the surrender of its "assigned amount" (carbon credits) at a 1.3-to-1-ton ratio for every ton of exceedance in a second commitment period. The use of this mechanism assumes that a second commitment period occurs, a detail not finalized at the time of the writing of the Bonn Agreement.

Third, the Bonn Agreements stipulated new funding mechanisms for developing countries. Perhaps the most novel form of funding under the Bonn Agreements is the decision to impose a 2% charge on Clean Development Mechanism. States agreed to use this surcharge to raise monies for the Adaptation Fund. Additional funds created include the Special Climate Change Fund that provides funds for adaptation, technology transfer, and economic diversification. The Least Developing Countries Fund assists with creating National Adaptation Programs of Action. Torvanger [36] notes that countries did not pledge specific amounts, except for the surcharge on the Clean Development Mechanism.

Finally, major elements of the Kyoto mechanisms needed further elaboration, including the concepts of fungibility (interchangeability), supplementarity, "hot air," and additionality. With the departure of the United States from the Kyoto Protocol, but not the negotiating sessions, the EU successfully politically isolated the United States from controlling the outcomes of the UNFCCC meetings. Consequently, the United States preferred policies for implementing the Kyoto mechanisms also lost salience. This loss meant that the EU could craft the carbon markets to match European moral preferences and political realities, i.e., ensuring that a corporation's ability to create carbon credits did not result in rewards for damaging the environment. The critical issue within this program development was not creating a new financial instrument but rather the moral question to what extent, if any, should corporations be allowed to create a financial reward for damaging the environment.

The parties agreed that all forms of the Kyoto mechanisms should be interchangeable. In other words, 1 ton of carbon dioxide equivalents from the emission trading scheme is completely compatible with a carbon dioxide equivalent generated through a Joint Implementation or Clean Development Mechanism project. Supplementarity means that states should comply with the emission reduction targets by adjusting domestic policies before utilizing global emission trading schemes. In practice, all Annex I countries use a combination of domestic policies and external trades to meet their emission reduction obligations. Last, hot air refers to the fact that the former Soviet Union, including Russia, experienced a severe economic downturn that caused an overallocation of assigned amounts to all of these countries. Allowing the hot air into the emission trading scheme with full fungibility would further weaken the environmental effectiveness of the Kyoto Protocol as other Annex I countries would be able to buy their way out of making carbon reductions.

Consequently, the conceptualization of "additionality" entered the climate change vocabulary. Additionality sought to ensure that any action taken under the Clean Development Mechanism would be in addition to regular business practice. The creation of additionality attempted to ensure that polluting entities do not profit from cleaning up pollution. In other words, project proponents (regardless of whether this entity is a state, a corporation, an organization, or an individual) would not be allowed to generate wealth through the Kyoto mechanisms for creating a reduction that would otherwise have occurred. For example, a landfill in a country that was required to install a methane collection system to prevent odor would not receive carbon credits for this action because the system meets a regulatory purpose. In other words, the installation of the methane collection system fails the additionality test as the business activity would have been taken anyway.

Wirth [43] described the Marrakesh Accord as a "unified set of rules" to implement the Kyoto Protocol. In essence, the Marrakesh Accord formalized the Bonn Agreements, with few exceptions [30]. Observers and scholars alike note that the Marrakesh Accords take on some of the attributes of an economic treaty in terms of its length and level of detail. This is perhaps, unsurprising, as the Marrakesh Accord essentially creates a new financial product—a carbon credit.

13.2   The European Union's Emission Trading Scheme

Thus, the EU created an internal emission trading program linked to the Clean Development Mechanism. Implementing the EU's emission Trading Scheme gave the EU the unique ability to control global emission markets through the concept of reciprocity. Trotignon [37] points out that this program serves as the world's first multinational carbon market. Therefore, it is the de facto model for other programs. Today, the EU Emission Trading Scheme remains the largest carbon credit market in the world [19].

The EU adopted its scheme in 2003, with its first phase occurring from 2005 to 2007. While the EU initially sought a carbon tax for emitters, internal European rules required unanimous consent to adopt the tax, and the measure failed in the 1990s [17]. Consequently, the EU turned instead to designing an emission trading market using the cap and trade model from the United States Acid Rain Program as inspiration. Initial design features required electricity generators and other industrial sources that emitted carbon dioxide to hold 1 European Union Allowance for every ton of carbon dioxide equivalent emitted. Sources that did not use all their allowances would be allowed to sell them to another party, while sources that emitted more than their allocation would need to purchase additional allowances. Impacted entities also had the option to reduce their emissions to align with their allocation.

The EU Emission Trading Scheme continues to change slightly over time as two additional phases occurred. The second phase took place from 2008 to 2012 and served as the cornerstone for EU compliance during the First Commitment Period as specified under the Kyoto Protocol. The third phase of the EU Emission Trading Scheme recently began in 2013 and ended in 2020. At the beginning of January 2021, regulated entities entered Phase 4 of the cap and trade scheme that was explicitly designed to meet the EU's commitments to the Paris Agreements.

Domestic emitters of greenhouse gases were allowed to utilize Clean Development Mechanism credits as an equivalent to the EU Allowance under certain circumstances. Clean Development Mechanism projects with known problems such as hot air or fraudulent conduct that do not meet the EU's standards are simply not allowed to count for domestic compliance purposes. As a result, the carbon credit concept expanded quickly worldwide as the G-77 at long last received the desired outcome—partners willing to make commercial investments in their countries in return for ownership of the carbon credits.

Meanwhile, in the United States, corporations that could potentially create carbon credits for a profit keenly felt the absence of the nascent carbon credit markets. While a mandatory market certainly stimulated demand, corporations also established a voluntary carbon credit market by creating supply. Indeed, multiple groups put forward plans for voluntary carbon credits, including the Gold Standard (2003), the American Carbon Registry American Carbon Registry (1996), and Verra's Verified Carbon Standard (2005). Buyer corporations found that carbon credits could aid in green advertising, including claims of carbon neutrality. Today, corporations outside of a mandatory program may utilize a multitude of voluntary programs to create carbon credits, keeping the Kyoto mechanisms near the forefront of carbon reduction strategies. However, these voluntary credits may not be used for compliance purposes in the EU Emission Trading System.

13.3 "Son of Kyoto"

As Annex I countries began to announce their positions on ratification, diplomats understandably turned their attention to operationalizing the Kyoto Protocol and focusing on the terms and conditions for a second commitment period. Two primary issues appeared during this time frame. First, the Kyoto Protocol itself had not passed the ratification threshold. Second, countries had not progressed in the strenuous task of negotiating a second commitment period, including the all-important details of what countries will be included in mandatory emission reductions and by how much.

After the United States announced its withdrawal from the Kyoto Protocol, states such as Japan and Russia took on increased importance as the ratification threshold included 55 countries representing 55% of the total carbon dioxide emissions from Annex I countries in 1990. While there was little doubt that 55 countries would ratify the Kyoto Protocol, serious concerns emerged about meeting the carbon dioxide emissions threshold as the United States represented 36% of the inventory.

Tiberghie and Schreurs [35] analyzed the rationale behind Japan's decision to ratify the Kyoto Protocol. They concluded that traditional power politics analysis fails to explain the rationale for Japan's ratification of the Kyoto Protocol. Japan potentially hurt its relationship with the United States and put its corporations at a competitive disadvantage vis-à-vis the United States and other regionally important economic competitors to do so. However, the Kyoto Protocol enjoyed significant public support. Additionally, part of the Japanese global identity includes international leadership on economic issues, and the main climate change agreement bore the name of a prominent Japanese city. While this embedded symbolism might seem a mere reason to ratify a treaty, maintaining international prestige does serve as a rationale for some political decisions.

With Japan depositing its ratification with the UN, global attention turned to Russia as it became the last country able to block the ratification of the climate change treaty. The Kyoto Protocol's impacts on the Russian domestic economy were complex by any standard. Russia potentially benefitted from the ability to sell carbon credits. Recalling from Chap. 8, the so-called "hot air" issue that stemmed from the collapse of the former Soviet Union meant that Russia's initial allocation was based on an economic activity level that no longer existed [7, 31, 40]. Thus, Russia had the ability to sell some of its "hot air" for profit.

However, the question of Russian ratification is, in reality, much more complex than its ability to sell excess carbon credits. Russia is a major source of natural gas on the world market, and the EU is one of its major clients. Consequently, ratification of the Kyoto Protocol brings with it a potential decline in future natural gas sales as all countries would have an obligation to reduce their reliance on fossil fuels. This simplistic analysis relies on the assumption that the EU would fail to take action in the face of a collapse of the Kyoto Protocol. However, the validity of this assumption will never be known, as Russia agreed to ratify the Kyoto Protocol in November 2004.

Russia did so, in significant part due to the EU's pressure on Russia in other negotiating forums, in particular the WTO. At an EU summit in May 2004, the EU and Russia worked out an agreement where the EU would not oppose Russia joining the WTO in exchange for the Russian ratification of the Kyoto Protocol [3]. Joining the WTO benefits Russia as it will gain access to other countries' markets under the principle of non-discrimination, meaning that the lowest possible tariff that applies to any one country is used for all. As a result of the Russian ratification, the Kyoto Protocol entered into force on February 16, 2005.

The global community did not wait to know the ratification fate of the Kyoto Protocol. Diplomats continued to meet to hammer out the details surrounding the so-called second commitment period. Perhaps the largest item of concern about the second commitment period involved the willingness of the developing countries to make binding commitments during this time frame. From an environmental stand-point, further reductions from developed countries would not be enough to limit potentially catastrophic climate change without meaningful and significant partici-pation from key developing countries. In fact, the key theme of the next round of negotiations in this arena could easily be characterized as the domination of the North–South divide, with the South vigorously defending its national interests.

While the diplomatic negotiations occurring as part of the climate change arena grew increasingly complex, this text will review two issues: the negotiations surrounding whether developing countries should sign on to mandatory mitigation actions (emission reductions) and, secondly, the question of funding to developing countries. This topic perhaps represents the sharpest difference between North and South, although the North's willingness to provide additional funds for adaptation to climate change and mitigation of greenhouse gases also demonstrates a stark difference of opinion.

Recalling from Chap. 8 the concept of common but differentiated responsibilities, developing countries emphasized that developed countries (Annex I countries in terms of the Kyoto Protocol) should lead the way by committing to steep targets before developing countries (non-Annex I countries) took action. The EU clearly viewed their targets in the first commitment period as meeting this criterion [18].

Greenhouse gas emissions inventories certainly support this stance. In 2006, China surpassed the United States globally as the single largest emitter of greenhouse gasses [20]. Le Quéré [32] ranked the top twenty-five carbon emitters in the world in 2017. The list placed five developing countries in the top ten, including China in first place and India in third. This ranking suggests that any further attempts to limit climate emissions without meaningful reductions from some developing countries will not yield sufficient results to stabilize carbon dioxide emissions in the atmosphere.

However, the G-77 response to this conundrum left much to be desired from an environmental perspective. Developing countries saw the expectation as an attempt to renege on the common but differentiated responsibilities principle and reiterated their stance that economic development triumphed over environmental protectionism. The United States quickly supported these developing countries' stance [24]. Thus, an early signal that coming to an agreement on a second commitment period would be neither quick nor decisive emerged from COP 8 in New Delhi, India.

After COP 8, the climate change negotiations moved forward but failed to provide meaningful breakthroughs or shifts in the dynamics that caused the North–South deadlock. Officially, negotiations on the second commitment period launched at COP 11/CMP 1 in Montreal, Canada, a meeting that gained notoriety as the first COP serving as the MOP under the Kyoto Protocol in 2005. However, the follow-up meeting COP 12 in Nairobi, Kenya, in 2006 yielded no meaningful progress on the conversations regarding future emission reductions.

13.4 The Bali Road Map to Copenhagen

It was left, then to Nusa Dua, Bali, Indonesia, to move the conversation forward at COP 13 in 2008. International diplomats continued to meet to discuss the path forward on climate change, with the objective of establishing mitigation targets in the second commitment period from 2013 to 2020. The work product of this session, the Bali Road Map, focused attention on five primary areas, including shared vision, mitigation, adaptation, technology transfer, and financing [38].

The Bali Road Map, in the words of Christoff [9: 472] is a "rough and narrow goat-track" rather than a well-paved road toward a second commitment period. At the end of the Bush Administration, the United States' reengagement within climate diplomacy, was simply put, unknown. The United States was due to elect a new president in 2008, and the only certainty was that President Bush was ineligible to run for a third term. Consequently, other countries wanted to lay the groundwork for welcoming the United States back into the treaty, while the United States sought to slow down the speed of the negotiations by insisting on developing countries' mandatory reduction targets.

The prospects of the United States returning to a leadership role improved dramatically with the Supreme Court of the United States ordered EPA to address greenhouse gases [27]. In this case, the state of Massachusetts sued the United States government for its failure to issue a ruling on whether greenhouse gases from automobiles endanger public health and welfare in the United States. If so, these gases constitute an air pollutant that must be regulated under the Clean Air Act, the primary air pollution control law within the United States. The Supreme Court sided with Massachusetts, triggering a series of events where EPA issued an endangerment finding that caused regulations on carbon dioxide and other greenhouse gases to be put in place as part of the United States regulatory system.

This one court case changed the logic of the United States negotiating stance by changing the economic principles underlying the negotiating position. Given that the Clean Air Act would soon force the United States-based corporations to make carbon reductions, requiring other countries' industries to make the same commitment would be one of the few ways to level out production costs.

While climate diplomats were undoubtedly heartened by the United States election of Barak Obama in 2008, current events emboldened climate skeptics. In 2009, climate skeptics hacked the email account of Dr. Phil Jones of the University of East Anglia in Norwich, England. The group found emails that implied that Dr. Jones and other climate modelers were intentionally misrepresenting data to show climate warming when, in fact, the data did not warrant it. The so called "climategate" scandal severely damaged the reputation of the International Panel on Climate Change [22]. More importantly, the scandal created an opening for climate skeptics to attack the entire process. This new scandal occurred just a few weeks before climate diplomats met in Copenhagen for COP 15 in 2009.

Once at Copenhagen, states disagreed on which countries should make targets and what the targets should be. Consequently, many states took on conditional targets that were dependent on the size and timing of other country commitments. In essence, the major states made their future commitments contingent on the participation of the United States. There is both an environmental and economic logic to this stance. Achieving any kind of atmospheric stabilization of greenhouse gases without the participation of the second largest emitter of greenhouse gases dooms the treaty to environmental failure. Further, the United States already enjoys significant economic advantages due to the abundant and easily obtained fossil fuel deposits. The United States, for its part, made their commitments contingent on China, the leader of greenhouse gas emissions currently, but not historically. The negotiations quickly stalled due to China's failure to commit to a mandatory greenhouse gas emission reduction pledge.

As a result, trust became an issue in Copenhagen [22]. Unsurprisingly, the twin tracks previously established refused to make meaningful commitments unless the other group went first. The developed countries negotiated new commitments through the Ad Hoc Group on Further Commitments from Annex I Parties track. They insisted on seeing the developing countries' commitments negotiated in the Ad Hoc Working Group on the Long-term Cooperative Action track. Consequently, neither group put forth their best efforts as each could potentially gain commercial advantage by being the last to commit.

Disappointingly, states spent time pointing fingers at other states rather than working together to resolve their differences. The United States certainly suffered from this viewpoint as other states and NGOs pointed out their failure to lead by calling them a laggard [10]. Equally importantly, the BASIC group consisting of Brazil, South Africa, India, and China, were called out for their lack of reduction commitments despite the BASIC group's actions in supporting other developing countries [15]. All the BASIC countries have rising greenhouse gas emissions levels. China produces the most greenhouse gas emissions in absolute terms, but not per capita (by population).

This domino status obscures another key facet of negotiating future commitments. States changed the baseline year to adjust the difficulty of meeting the targets as states' greenhouse gas emission levels have increased since 1990. However, the quality of the greenhouse gas emissions inventory has improved since that date, potentially increasing the accuracy with respect to the quantity of greenhouse gases in the economy and the atmosphere.

The Copenhagen Accord, a political agreement drafted directly by world leaders attempted to gloss over these differences. The two-page document contained very little in terms of mandatory new commitments to reduce greenhouse gases in the future. Upon its completion, President Obama and others rushed to hold a news conference announcing the agreement offending countries that had not been updated on the agreement. States officially took "note" of the Copenhagen Accord, a phrase used to disguise states' failure to agree to a meaningful outcome. The second commitment period from 2013 to 2020, with its emphasis on binding emission reductions for Annex I countries, did not emerge. Accordingly, EU leadership on this issue also diminished, with the United States resuming some of its former leadership role under the Obama administration.

Dimitrov [15] blames three countries for this failure—China, India, and Brazil. Grubb [21] disagreed, pointing instead to the EU's capitulation to the United States' demand for a treaty-based upon domestic actions, the so-called bottom-up approach, without receiving anything in return. This approach effectively meant that the United States will align its climate change mitigation strategies with domestic priorities rather than negotiating international agreements that would be imposed from the top-down and might not be in the best economic interest of the United States.

More damagingly, the combined emission reductions pledged at Copenhagen do not accomplish the overall goal of limiting the rise of global temperatures to 2 °C [12]. While the greenhouse gas emission targets represent a net decrease in emissions, more reductions would be needed to close the gap. Further actions on climate change would require concentrated efforts not only from major emitting states like China and India but also from non-state actors that also contribute to governance either through direct control of emission sources or through their ability to encourage deeper cuts by those who do.

13.5   From Copenhagen to Paris

The aftermath of the collapse of negotiations around a second commitment period was not the end of climate change diplomacy. A myriad of activities erupted as states, international organizations, substate actors, and non-state actors sought ways to achieve new emission reductions outside of the formal negotiating sessions. Within the UNFCCC negotiating processes, two conferences stand out as key change moments for the rapidly evolving climate change diplomacy, including COP 17 in Durban, South Africa, that led to the Paris Agreement at COP 21.

COP 17/CMP 7 in Durban established 2015 as the deadline for completing a new international arrangement to continue the progress made toward a global climate change regime under the Kyoto Protocol. Countries agreed to form the Ad Hoc Working Group on the Durban Platform for Enhanced Action. This working group negotiated a new draft negotiation text that would become the basis for the Paris Agreement over the course of the next several years.

Interestingly, the Durban outcomes also recognized the fragility of the state's commitment to using international diplomacy as the sole mechanism to reduce greenhouse gas emissions. Thus, the UNFCCC Secretariat sought to encourage non-state activity by launching key initiatives such as Momentum for Change. This program seeks to recognize actions by a variety of actors, including governments, businesses, and NGOs, that reduce climate change emissions while promoting other social changes such as gender equality, climate justice, and poverty eradication. Each year, the UNFCCC recognized organizations as Light-house Activities. Winning organizations and their projects are featured on a special website created for this purpose.

Second, the UNFCCC Secretariat, along with support from key states hosting the COP, created the Non-state Actor Zone for Climate Action (NAZCA). NAZCA consists of an online portal where non-state actors of any sub-type may register their support, including any actions taken to achieve climate reductions. As of February 2021, the portal makes notes of 18,556 actors recording 27,513 activities [39].

In light of the abysmal track record of the climate change negotiations from Marrakesh to Copenhagen and beyond, the turnaround at Paris during COP 21 truly represents a satisfactory outcome for the Annex I countries and a potential victory for the environment. For the first time, developing countries agreed to accept the need to reduce greenhouse gas emissions. All countries are committed to establishing a Nationally Determined Contribution for future greenhouse gas emission reductions.

Theoretically, each country would increase its emission reduction target for the next five-year period. Domestic policy approaches unique to each country combined with an international treaty requirement stating that countries should continue to reduce their greenhouse gas emissions. Further, each state's domestic policies, nationally determined contribution, and mandatory reports will be subject to international scrutiny.

The UN most emphatically heralds the Paris Agreement as a legally binding treaty; this may, in fact, not be the case due to the wording of the agreement. Legal scholars within the United States point out that the Paris Agreement relies on the word "should" and, as such, did not create an obligation by a state, but only an inspirational goal [6, 16]. Further, Presidents Obama and Biden pledged to adhere to the agreement without Senate ratification. Senate confirmation is a constitutional requirement for the United States to be bound by a treaty. Other countries disagree. Notably, the EU strongly supports aspirational goals and treaties possessing the same status as enforceable law. As such, many other countries consider the Paris Agreement to be binding international law.

While it is too early to tell if the Paris Agreement will successfully reduce greenhouse gas emissions globally, the international process certainly brought hope to people around the globe that the threat of climate change would eventually be brought under control. However, the great promise of a global climate regime quickly ended as the world realized that President Obama would be replaced with Donald Trump, a militant opponent of the Paris Agreement.

President Trump entered office in January 2017 with significant domestic opposition to his election. On June 1, 2017, President Trump announced the United States withdrawal from the Paris Agreement. However, UN rules written into the Paris Agreement effectively prevented the United States from formally acting on this intention until November 2020. The United States withdrawal from the Paris Agreement signaled yet another defeat for the international ambition for a worldwide climate change agreement. At the same time, existing domestic requirements for climate change reductions remained largely untouched, bolstered by the same domestic laws that required the initial regulation of greenhouse gases in 2008, the Clean Air Act. At the writing of this manuscript, it is too early to conclude whether the Paris Agreement may yet live up to the rhetoric proclaiming its arrival in 2015.

13.6 Climate Scholarship

The continued evolution of climate diplomacy has at once frustrated and stimulated scholars. New conceptualizations of the climate change arena emerged as the shape of the negotiations, and their outcomes changed. During the early years of the Kyoto Protocol, scholars framed this arena as a climate change regime dominated by states using a top-down approach to emission targets, reduction strategies, and timetables [4, 13, 28, 44]. Thus, this section begins by reviewing the functional design of the climate change agreements before examining scholarship around regime theory.

International relations scholars have long recognized that varying activity levels interact with one another. In plain language, the international realm of states may shape domestic outcomes [14]. Individual states would then ensure that everyone followed the domestic laws and regulations implementing the main international treaty. For example, a state may commit to lowering domestic emissions by 6% overall. The state would then raise fuel economy standards on cars in combination with adding new natural gas fired power plants to replace old coal fired power plants.

While hope continued that the United States would rejoin this top-down approach to global climate policy when President Bush left office, the environmental realities of enacting an environmentally effective regime absent the participation of China, the world's largest (and rapidly growing) carbon emitter, ended the usefulness of the top-down approach. Very few options were left structurally; only a bottom-up approach centered around domestic actions and a hybrid option that combined the bottom-up and top-down approaches.

Thus, the top-down process from international to national may not hold true in every circumstance. In this case, domestic events within the United States, namely the Supreme Court of the United States' opinion in *Massachusetts v. EPA* [27], ruled that the EPA could not avoid regulating greenhouse gases under existing domestic law (the Clean Air Act). The Supreme Court ruling benefitted the global negotiations by incentivizing the United States to return to the international negotiating table if only to protect its industrial economic advantage. Simply put, changes in the domestic laws and regulations allowed a change in negotiating stance at the international level, a textbook example of DeSombre's theory that states domestic preferences may shape international outcomes [14].

More recently, scholars have described climate change diplomacy as creating a hybrid multilateralist system that creates a hybrid of the top-down and bottom-up approaches [5]. On the one hand, the Paris Agreement requests that states implement domestic policies to lower their domestic greenhouse gases. On the other hand, the UNFCCC Secretariat includes a bottom-up element that promotes carbon reductions by non-state actors through the NAZCA. Additionally, de Oliveira [11] pointed out that subnational governments make significant contributions to reducing climate change. Gupta [22] points out that the contributions of subnational governments, including cities, are to "observe" the negotiations, but more importantly, to implement policies that create emission reductions. In a similar vein, Dimitrov [15] suggests that the most significant actions that contribute to emission reductions are taken by non-state actors away from the formal conference negotiations.

Initially scholars believed that a climate change regime formed with the ratification of the Kyoto Protocol. Scholars certainly expected in the aftermath of the Kyoto Protocol that a climate change regime had formed and that states would continue to expand on the rules for carbon trading necessary to create a workable global market [4, 13, 28, 44]. Instead, the United States executed a rarely used withdrawal from the Kyoto Protocol and, in the process, stopped the regime from forming as consistent behaviors did not emerge. While the time period between the end of the Kyoto Protocol in 1997 through the Marrakesh Accords could be analyzed as the establishment of a regime, more recent events lead scholars to believe that actors' expectations did not converge around a single institution. Haas [23] noted that international environmental diplomacy is rarely straightforward and that the failure to agree in the past does not mean that diplomats will not agree in the future.

That is not to say that the failure of the climate change regime to form negated the entire body of scholarly research on regimes. Instead, it should be seen as an example of a particularly long and difficult negotiation where global actors may devise yet another new structural relationship in the long history of international affairs. As climate change negotiations stalled out, proponents of strong environmental controls increasingly recognized successful activity to reduce carbon emissions even if the actions were technically not required by the various protocols, treaties, and agreements that make up climate change diplomacy.

In contrast, to a singular regime, regime complexes occur when two or more functionally independent institutions manage the same partially overlapping issue area [1, 2, 26, 42]. Recalling from Chap. 5 that the ozone depletion regime began taking action under the ozone rules to reduce greenhouse gases, two or more regulatory schemes for climate change now exist. Further, voluntary initiatives around the world also serve as functionally independent institutions, albeit one without the legal authority of the UN.

Scholars concluded that the regime complex relationship might be more able to withstand environmental and political uncertainty than a singular regime [1, 26]. This structure consists of horizontal movements at the state level and horizontal movements that may connect a national government to a subnational government. A failure to act at one point may be compensated by action at a different point. In other words, if a state stalls out negotiations in one negotiating forum, a shift to a second negotiating forum may allow other willing parties to continue to move forward. As we have seen in the past, a lack of forward progress in climate diplomacy may well be taken up by other regimes, such as ozone. Equally important, states may work directly with subnational governments to create a meaningful exchange of ideas and carbon reductions.

Upon concluding our examination of the most recent history of the climate change regime complex, the sheer number of institutions, organizations, and individuals taking meaningful actions is impressive. They are not only working to reduce their carbon emissions but are participating in a wide variety of events to encourage (or force) others to do likewise, thus decreasing the likelihood that significant damage occurs to our ecosystem in the future. These groups are undoubtedly more optimistic about the future with the inauguration of President Biden, given his recent pronouncement that the United States will rejoin the Paris Agreements [25].

References

1. Abbott KW (2012) The transnational regime complex for climate change. Environ Plann C 30(4):571–590
2. Abbott KW (2014) Strengthening the transnational regime complex for climate change. TEL 3:57
3. Baker P (2004) Russia backs Kyoto to get on path to join WTO. WaPo May 22
4. Bodansky D (2001) The history of the global climate change regime. Int Relat Glob Clim Change 23(23):505
5. Bodansky DM et al (2016) Facilitating linkage of climate policies through the Paris outcome. Clim Policy 16(8):956–972
6. Bodansky D (2015) Reflections on the Paris conference. Opinio Juris 15
7. Bohringer C (2000) Cooling down hot air: a global CGE analysis of post-Kyoto car-bon abatement strategies. Energ Policy 28:779–789
8. Brouns B, Santarius T (2001) Die Kyoto-Reduktionsziele nach den Bonner Beschlü-ssen. Energiewirtschaftliche Tagesfragen 51(9):590–591
9. Christoff P (2008) The Bali roadmap: climate change, COP 13 and beyond. Environ Polit 17(3):466–472
10. Christoff P (2013) Cold climate in Copenhagen: China and the United States. Environ Politics 19(4):637–656
11. de Oliveira JAP (2009) The implementation of climate change related policies at the subnational level: an analysis of three countries. Habitat Int 33(3):253–259
12. Dellink R, Corfee-Morlot J (2010) Costs and effectiveness of the Copenhagen pledges: assessing global greenhouse gas emissions targets and actions for 2020. OECD, Paris
13. Depledge J (2013) The organization of global negotiations: constructing the climate change regime. Earthscan, New York
14. DeSombre ER (2000) Domestic sources of international environmental policy: industry, environmentalists, and US power. MIT Press, Cambridge

15. Dimitrov RS (2010) Inside UN climate change negotiations: The Copenhagen conference. Rev Policy Res 27(6):795–821

16. Dimitrov RS (2016) The Paris agreement on climate change: behind closed doors. Glob Environ Polit 16(3):1–11

17. Ellerman AD (2010) The EU emission trading scheme: a prototype global system? Post-Kyoto international climate policy: implementing architectures for agreement. Cambridge University Press, Cambridge, pp 88–118

18. EU (2003) COP 9/climate change: all parties must maintain momentum to tackle the 21st century's biggest environmental challenge. IP/03/1638

19. EU (2021) EU emission trading system (EU ETS). https://ec.europa.eu/clima/policies/ets_en. Accessed 17 Jan 2021

20. Frohlich TC, Blossom L (2019) These countries produce the most CO_2 emissions. USA Today. https://www.usatoday.com/story/money/2019/07/14/chinaus-countries-that-pro duce-the-most-co-2-emissions/39548763/. Accessed 28 Jan 2022

21. Grubb M (2010) Copenhagen: back to the future? Ed Clim Policy 10(2):127–130

22. Gupta J (2010) A history of international climate change policy. Wires Clim Change 1(5):636–653

23. Haas PM (2008) Climate change governance after Bali. Global Environ Polit 8(3):1–7

24. Jacob T (2003) Reflections on Delhi. Clim Policy 3(1):103–106

25. Kann D, Atwood K (2021) Paris climate accord: Biden announces US will rejoin landmark agreement. CNN https://www.cnn.com/2021/01/20/politics/paris-climate-biden/index.html. Accessed 31 Jan 2022

26. Keohane RO, Victor DG (2011) The regime complex for climate change. Perspect Polit 7–23

27. *Massachusetts v. EPA* (2007) 549 U.S. 497

28. Oberthür S (2011) The European Union's performance in the international climate change regime. J Eur Integr 33(6):667–682

29. Ott HE (2001) The Bonn Agreement to the Kyoto Protocol–paving the way for ratification. Int Environ Agreem 1(4):469–476

30. Ott HE (2002) Climate policy after the Marrakesh accords: from legislation to implementation. Wuppertal Institute for Climate, Environmental and Energy Wuppertal, Germany

31. Paltsev SV (2000) The Kyoto Protocol: "hot air" for Russia? Department of Economics, University of Colorado Working Paper No. 00-9 October

32. Le Quéré G et al (2018) Global carbon budget 2018. Earth Syst Sci Data 10:2141–2194

33. Risen J (2010) Veterans sound alarm over burn-pit exposure. NYTimes. https://www.nytimes.com/2010/08/07/us/07burn.html. Accessed 9 Jan 2021

34. Sidel VW (2008) War, terrorism and the public's health. Med Confl Survival 24(S1):S13–S25

35. Tiberghie Y, Schreurs MA (2007) High noon in Japan: embedded symbolism and post-2001 Kyoto Protocol politics. Global Environ Polit 7(4):70–91

36. Torvanger A (2001) An analysis of the Bonn Agreement: background information for evaluating business implications. Report/CICERO-Senter for klimaforskning. http://urn.nb.no/URN:NBN:no-3645

37. Trotignon R (2012) Combining cap-and-trade with offsets: lessons from the EU-ETS. Clim 12(3):273–287

38. UNFCCC (2021a) Bali road map intro. https://unfccc.int/process/conferences/the-big-picture/milestones/bali-road-map. Accessed 10 Jan 2021

39. UNFCCC (2021b) Global climate action NAZCA. https://climateaction.unfccc.int/. Accessed 6 Feb 2021

40. Victor DG, Nakicenovic N, Victor N (1998) The Kyoto Protocol carbon bubble: implications for Russia, Ukraine, and emission trading. IIASA Interim Report IR-98-094

41. Vrolijk C (2001) COP-6 Collapse or 'to be Continued …?' Int Aff 77(1):163–169

42. Widerberg O, Pattberg P (2017) Accountability challenges in the transnational regime complex for climate change. Rev Policy Res 34(1):68–87

43. Wirth D (2002) The sixth session (part two) and seventh session of the conference of the parties to the framework convention on climate change. Am J Int Law 96(3):648–660
44. Yamin F, Depledge J (2004) The international climate change regime: a guide to rules, institutions and procedures. Cambridge University Press, Cambridge

Chapter 14
Transforming the World Through the 2030 ASD

Abstract As the Millennium Development Goals time frame ended, states recognized the usefulness of this program and moved to create a successor to assist in implementing sustainable development. Gathering at Rio de Janeiro, Brazil, on the twentieth anniversary of the Earth Summit, states failed to create new momentum for international environmental diplomacy, with the exception of creating *The Future We Want*, a call for states to maintain their focus on sustainable development. States exercised their privileged position within the international system to control the process of creating the new program. Consequently, the 2030 Agenda for Sustainable Development expanded the Millennium Development Goals. The United Nations responded to criticisms of its management of the previous program by strengthening its capacity to manage data and to promote a program that relies on both state and non-state actor participation. This chapter also examines the actor networks that seek to implement sustainable development goals as recorded in the Partnership Data for Sustainable Development Goals database.

Keywords 2030 Agenda for Sustainable Development · Partnership Data for Sustainable Development Goals · Sustainable development · Networks · Rio + 20 · Sustainability indicators

The great blessing, or occasionally the great curse, of international environmental diplomacy, is the unwillingness to end programs; diplomats have incentives to avoid publicizing failure [2]. Each new chapter within international environmental diplomacy begins where the last issue area left off. When there is an explicit predecessor agreement, continuity of principles, ideas, and modalities occurs more often than not. Thus, the 2030 ASD continues the work of the MDGs, as discussed in Chap. 12, to relieve poverty across the globe.

Recalling also from Chap. 12, one of the lasting critiques of the MDGs lambasted the technocratic (rather than diplomatic) process. While the idea for the 2030 ASD carried on the MDGs, the UN system returned to the established negotiating practices that constitute international environmental diplomacy to create the 2030 ASD. While this could have launched protracted negotiations about the shape of the post-MDG regime, it did not. Instead, the 2030 ASD arose from an inauspicious beginning as diplomats launched negotiations toward a new set of benchmarks as one of the few appreciated developments from the United Nations Conference on Sustainable Development (Rio + 20).

In an international diplomatic rarity, participants and observers alike did not hesitate to declare the Rio + 20 Conference that marked the Earth Summit's twentieth anniversary as a failure [15]. Returning to Rio de Janeiro, Brazil, in 2012, global diplomats assembled from June 20–22, 2012, with the third preparatory session occurring immediately before the conference. Unfortunately, the conference produced few memorable moments. By the end of the three-day meeting, the conclusion that international environmental diplomacy had reached an all-time low was inescapable. The agreement between member states to negotiate what would become the 2030 ASD stood as the sole promising work of this symbolic global environmental politics milestone.

This chapter reviews the 2030 ASD, beginning with its origins as part of the failed Rio + 20 conference. Like its predecessor program, the 2030 ASD functions as a series of benchmarks to measure countries' developmental and environmental status. It is an agreement where all national governments commit to achieve the same global goals and to measure these goals in the same way. However, this agreement does not revolve around one environmental problem. Many of the previous issue areas highlighted in this book took the form of a regime in the sense that diplomats created treaties based on one problem like the ozone hole and committed to working together to close the hole. Instead, the SDGs put forth a normative agenda for all international societies to adopt.

Like the MDGs before it, the 2030 ASD created the SDGs as a global agenda for the next fifteen years. Consisting of seventeen goals and 169 targets, with 303 indicators, the SDGs represent an ambitious attempt to guide global policy in a common direction. It does so by collecting country-specific data in order to raise awareness of the status of each country. This awareness may facilitate additional resources or changes in policies that contribute to the betterment of global society. The national level statistical analysis does not allow for an opt-out provision. In this way, the statistical analysis treats an NGO sponsored project the same as a government sponsored project. In other words, the data collected to monitor the 2030 ASD pulls together all the efforts of governments, non-profits, and business and industry groups that impact society. Importantly, the benchmarks do not account for the motivation of the group seeking to "improve" society. Contributions from groups that might have hesitated to associate themselves with a UN led initiative will nevertheless be captured in the data.

The size and the complexity of this agenda make it much more difficult for any group, whether that group consists of states, international organizations, non-state actors, business and industry groups, or some combination of partnerships, to exercise control over these items. While states may have responsibility for ensuring the welfare of their citizens, they may not be the sole actor that enables (or prevents) the achievement of the SDG domestically. Regardless of the nature of the 2030 ASD, this benchmark nevertheless represents honest efforts to improve the lives of millions of people around the globe.

14.1 Transitioning from the MDGs to the SDGs

Given the success of the MDGs in facilitating increased financial aid through both private and public channels, it is, perhaps, not surprising that the developing countries were eager to see the continuation of international development goals past the original 2015 deadline. The diplomatic discussions to create a post-2015 agenda began at the 2010 Millennium Development Summit in New York City on September 20–22. More formally known as the High-level Plenary Meeting of the UN GA, attendees focused primarily on intensifying efforts to meet the MDGs by 2015. However, the outcomes document for this meeting directed then Secretary-General Ban Ki-Moon to advise states on this matter [22]. In response, he appointed a UN System Task Team on the Post-2015 Agenda comprised of a variety of UN system members and independent international organizations. The UN DESA and the UNDP co-chaired this task force.

The inclusion of UNDP in a report within this genre is unsurprising given that much of the MDGs focus on development issues. UN DESA, however, resides as an office within the UN Secretariat. Created in 1948, UN DESA describes itself as a "think tank" for the UN system and the member states that created the organization [21]. UN DESA provides services by collecting, analyzing, and publicizing a variety of statistical analyses dealing with global problems, including sustainable development. Additionally, UN DESA also provides facilities for international conferences and assists countries with capacity building projects regarding data creation, data analysis, and data management tasks. Thus, member states heavily rely on UN DESA to carry out many administrative tasks associated with the 2030 ASD.

This group released its first report, *Realizing the Future We Want for All*, in June 2012. The report highlighted the need to retain focus on economic development and environmental sustainability while adding new goals dealing with social inequality, human rights, urban areas, and peace and security [24]. It also recognized the need to balance creating legitimacy through widespread participation with the advantages of limiting the number of participants creating new goals and targets in order to reach consensus.

States opened a parallel conversation as part of the Rio + 20 conference when Colombia and Guatemala proposed the concept of a new set of sustainable development benchmarks during the preparatory sessions for the Rio + 20 summit. While the proposal for the second set of benchmarks found widespread support among countries as part of the Rio + 20 Summit, the timing and negotiating processes for this global summit were less than ideal.

Pattberg and Mert [14] point toward global circumstances that negatively impacted the outcome of the Rio + 20 Summit, the Global Recession that lasted from 2008–2012, and the need to operate by consensus. At the Rio + 20 conference, the Great Recession of 2008–2012 and the EuroZone debt crisis meant that countries' economic growth noticeably declined. Individual unemployment rose, and inequalities between rich and poor increased within countries and between countries. Financial flows between developed North countries and developing South countries also dropped off.

As a result, the timing to ask for new funding or other forms of assistance came at a time when Northern countries were less inclined to send financial assistance overseas. Thus, politicians and diplomats faced an unwelcome dilemma between creating a controversial document where disagreements between countries would be center stage or agreeing to a short, non-descript document that omitted these differences. Brazil, as the host country, elected for the consensus approach. Consequently, the Rio + 20 outcomes, *The Future We Want*, appeared to be set for mediocrity.

Nevertheless, the international community broadly supported the idea of beginning conversations between states as part of the Rio + 20 conference outcomes. This group convened along with the UN System Task Team already at work under the oversight of the UN Secretary-General. Instead, member states created the Open Working Group (OWG) and limited its membership to 30-member states based on the five regional groups rather than on common interests or the traditional power groupings of the G-77 and China, the European Union, the United States and its allies, and Russia and the other economies in transition states.

However, only six of these "seats" were held by individual countries. Nine pairs of countries shared nine seats, while fourteen groups of three accounted for fourteen seats. Four countries, Algeria, Egypt, Tunisia, and Morocco, shared the last seat. As a result of this seat sharing initiative, individual countries reconciled their differences in order to interface with the larger group as a whole.

The UN GA tasked the OWG with creating a post-2015 agenda that reinforced sustainable development while being universally applicable, action-oriented, and easily understood [23]. This document expressed the admittedly utopian viewpoint of the UN in arguing for a peaceful, democratic, and sustainable world that allowed for economic growth according to the need of each country. At the same time, the document also paid homage to a long-standing principle within international politics, state sovereignty and its corollary, the principle of non-interference. By setting goals but leaving open the path by which the goals are achieved, the UN struck a remarkable balance between harmonizing norms, values, and outcomes, while refraining from dictating to other states how to act.

States met in thirteen sessions from March 2013 to July 2014. The OWG process, while formally dominated by states, included a broad representation of other interests, including viewpoints from within the UN as well as other international organizations. Private sector viewpoints (through relevant NGOs) along with other members of civil society and invited guests from academic circles also contributed to the OWG process.

The OWG process is split into two phases, with phase one reviewing issue areas and phase two crafting the final report for the UN GA. Phase one identified 19 potential issue areas that should be included within the SDGs as well as the cross-cutting themes that impact all the remaining goals. For example, partnerships for the goals outline how different groups could meaningfully contribute to the other issue areas. Additionally, participants in the OWG process also discussed principles of importance to the group. One of the areas of discussion during the OWG meeting included how to apply the principle of common but differentiated responsibilities to the SDG.

States also critiqued the donor-centric focus that traditionally dominated poverty eradication. Participants in the OWG process recognized the limits of international development assistance in providing all the financial resources needed for the transformation envisioned by the SDGs. Some developing countries believe that the developed countries pledged 0.7% of their gross national income to assist them in the development process.[1] However, the developed countries believe that it is their choice of how much money to spend and what to spend it on. Consequently, ODA from individual countries rarely meets the 0.7% threshold.

Developing states' critiques of ODA should not be seen as a rejection of this system but rather a calculated ploy to encourage developed states to donate more financing or to argue for the creation of additional mechanisms that supplement official channels. Thus, the crafting of the draft text deliberately created mechanisms by which to engage the private sector and the non-profit sector. The private sector also contributes funding to developing countries through their charitable giving. Corporations control considerable wealth, and many have global philanthropic activities in addition to their for-profit operations. Similarly, non-profit operations control billions of dollars in assets that may be used for philanthropic purposes. Consequently, the OWG also considered "means of implementation" when creating the document. Given the incredible disparity between countries and peoples, it is hardly surprising that this vital topic appears in multiple places within the document. Most significantly, the OWG included this topic in a standalone goal. Diplomats also wrote the means of implementation directly into targets for other goals.

Additionally, the OWG also provided ideas on the communication of the SDGs. Widely hailed as one of the strengths of the MDGs, member states sought to maintain this element in the post-2015 system. Ideas generated included creating country-specific dashboards for monitoring. The need for quality data is itself included as a target in the Zero Draft transmitted to the UN GA.

The OWG delivered its final product to the UN GA in time for its sixty-eighth session in 2014. The draft identified potential goals and targets but not indicators. The document does not propose creating new legally binding commitments, nor does it require the renegotiation of any existing treaty. Instead, the Zero Draft contains a list of interlinked issue areas that countries could focus on to improve the quality of life for their citizens.

The UN GA convened a second round of negotiations to transform the Zero Draft into its final form, *Transforming Our World, The 2030 Agenda for Sustainable Development*. This round of negotiations returned to the tradition of inviting all interested member states to participate in crafting the final draft document. Like the OWG before it, this intergovernmental negotiation also welcomed participation from a variety of voices that reflect global society. This work group convened eight times between January and July 2015.

The group tasked the UN Statistical Commission, a functional commission (subdivision) of the ECOSOC, with beginning work on the indicator portion of the SDGs.

---

[1] Conversations about a target for ODA date back to the Pearson Commission in 1969. The United States never formally agreed to this measure.

These global indicators would then be reviewed by ECOSOC and the UN GA for adoption by member states. These indicators may be supplemented by additional data sets from other sources.

States officially adopted the 2030 ASD at the UN Summit on Sustainable Development in September 2015, just in time for the beginning of the next 15-year development interval from January 1, 2016 to December 31, 2030. While countries and the UN system reacted favorably to the creation of the SDGs, scholarly reaction to the SDGs has been more mixed. Weber [25] argues that the form and structure of the SDGs inherently advance the neoliberal capitalist agenda at the expense of a more transformative shift to create an environmentally friendly and socially just society. In other words, the creation of goals to increase the amounts of ODA serves to reinforce the system that created the inequalities rather than shifting the current neoliberal system to eliminate the need for wealth transfer in the first place.

14.2   New Goals and Targets

The 2030 ASD continues the work of the MDGs that were in place for 2000–2015. While the MDGs constituted a broad look at ending poverty, the SDGs seek to achieve the same result by providing more specific goals and targets [26]. Chasek and Wagner [3] believe that the 2030 ASD breaks away from its past linkages to the MDGs while nevertheless providing a meaningful set of benchmarks for all countries in the quest for a more equitable world.

As the MDGs were not legally binding commitments but rather aspirational goals, there are no consequences to countries that failed to meet their targets. Equally important, countries that did meet their goals should be considered as having transitioned from developing countries to developed country status. Nor does meeting the MDG exempt a country from the current SDG process.

The 2030 ASD contained 17 goals for all countries to accomplish, along with 169 targets and 303 indicators. For example, Goal 14, Life below on water, contains ten targets that represent actions that should be undertaken to maintain a healthy marine environment. One of these targets, Target 14.1, encourages the elimination of marine pollution from land-based sources.

Like the MDGs, the SDGs contain aspirational goals rather than legally binding goals. Individual countries cannot be taken to court for failing to achieve these targets. Instead, the SDGs serve as an action guide that countries should undertake on behalf of their citizens. In this sense, the journey itself benefits the countries that strive to make life better for their citizens.

While developing countries intended for poverty reduction to be the centerpiece of the SDGs (and zero poverty is indeed the primary goal), overall, the SDGs represent a better balance between the three pillars of sustainability: environment, economic, and social. These goals represent attempts to end poverty and hunger, provide education that leads to fulfilling and meaningful work, improve the quality of urban areas, encourage wise use of natural resources, protect the environment, and ensure that all peoples are created with dignity. Recognizing that the inability of countries to meet these goals without assistance is itself an indicator of global inequality, partnerships for the goals emphasize the need to work together to end these wicked problems.

While the goals appear to be directed primarily at assisting the movement of developing countries toward sustainable development, the goals have also found wide-spread support in developed countries. In other words, the SDGs, while created as a means to end poverty and improve equality, are not just items of concern for the developing South. These ideals also encapsulate a moral authority for improving the conditions of peoples in the developed North that continues to see an increasing gap between rich and poor domestically. Thus, Northern countries also utilize the SDG framework as a public policy tool. As a result, work toward achieving the SDGs has been nearly universal both domestically and internationally.

The seventeen SDGs should not be seen as independent issue areas. The SDGs interact to create a complex issue network [11]. The nature of the interactions varies, with scholars classifying the interactions as positive or negative [12, 16]. Scholars tend to identify the interactions using slightly different methodologies, however, a typical stance notes that positive interactions amplify both methods in a desirable direction. Conversely, negative interactions interfere with each other. For example, increasing industrialization to create jobs will likely generate more greenhouse gases that exacerbate climate change impacts.

Targets take on both qualitative and quantitative forms. For example, Target 1.2 requires countries to reduce by 50% the number of people living below the extreme poverty line. This represents a quantitative goal with a relatively well-defined methodology for enumerating the number of impoverished citizens within a country. Alternatively, targets may take on a qualitative form, as is the case in Target 16.3. This target requests that countries promote the rule of law at the national and international levels.

The UN created the UN SDG Global Database to publicize the SDG indicator data. Despite creating a significant set of data indicators and the data warehousing necessary to compile and publicize the data, countries can and do decide to forgo reporting on every indicator. The absence of country-specific data for an indicator could also indicate that the data is available but not collected. More problematically, work on creating indicators has not caught up with the desire to use the data, so much so that the UN created a three-tiered process for assessing the quality of the indicator itself.

The three tiers vary in stringency, with Tier I consisting of indicators with strong linkages that are in use for at least half of the countries where the indicator is relevant. Tier II indicators differ from Tier I indicators in that the indicator exists but is not in widespread usage. Finally, Tier III indicators denote the absence of an international methodology or standard. As of March 2021, the UN SDG Global Database indicates that every SDG indicator falls into either Tier I or Tier II [20].

Perhaps in response to harsh criticisms regarding the availability and quality of real-time data under the MDGs, the UN reacted much more strongly in managing data related to the 2030 ASD. After all, data on the SDGs should be considered a public good in and of itself. Further, it is worth pointing out that some countries would not be able to generate the data used in the SDGs without some significant assistance in the form of capacity building and financial assistance.

Secretary-General Ki-moon emphasized the importance of data by creating the Independent Advisory Group on a Data Revolution for Sustainable Development. This group noted that the data revolution contributed to the North–South gap. The ability to participate in the data revolution and the benefits from using high quality data allowed the United States and Europe to benefit more than other parts of the world [10]. This workgroup concluded that improvements in data would be necessary to support the SDGs. They recommended concrete steps to improve data collection, including creating a hub for publicly available data within the UN, enhancing the capacity of developing countries to produce their own data, and ensuring that the benefits of data collection and utilization benefit all countries.

Secretary-General Ki-moon agreed to create new infrastructure within the UN and launched the Global Partnership for Sustainable Development Data, with membership spread across states, civil society, and corporations. This group supports the creation and management of data for various end users including countries, lenders, and other interested users. Adams [1] cynically speculates that this initiative may do more to promote and upgrade infrastructure for providing development aid and assistance than solving the underlying issues of poverty.

As a result of the emphasis on data collection in the SDGs, the UN significantly increased its support for data distribution related to the SDGs. One database of note, the Partnership Data for Sustainable Development Goals (PD4SDG), established an internet-based database where project proponents may register activities that further the SDGs. This database continues the trend of incorporating non-state actors into prominent positions within the global society. In fact, the project database itself barely distinguishes between states and non-state activity.

Egelston et al. [5] analyzed the relationships between the project proponents in the database by applying mathematical tools for analyzing social networks. The insights provided by this new methodology confirm the shift in social structure from a state-centric focus to a polycentric focus. In other words, while the SDGs have state-centered actions at the heart of the actions requested, the UN actively engages all actor types in the process of implementing the SDGs in accordance with the directives established in Goal 17.

UN DESA altered the PD4SDG database structure multiple times during its relatively short history. One of the more intriguing additions requested that project proponents declare that project criteria should be specific, measurable, achievable, resource-based, and time-based. However, the PD4SDG database has not historically rejected entries that do not meet these criteria [5].

Projects within the PD4SDG database reveal a tremendous variety of actions with regard to the SDGs. Activities range from simple statements pledging to support a more environmentally friendly position, such as Ramapo College of New Jersey pledging to act more in line with the UN SDG Initiative [18]. Other projects may represent more wide-reaching efforts as the state of Colombia completes its expansion of its marine protected areas with technical and financial assistance from a network of NGOs [19].

14.3 Critiques, Changes, and Challenges

Analysis about whether the SDG structure compares favorably to the MDGs deserves attention. Certainly, reviewing the SDG negotiation process and the changes within the international system stemming from the negotiation process is also worthwhile. When examined chronologically, each issue area that diplomats negotiated success- fully typically involved innovation and experimentation. Social learning occurred within the international environmental system, where diplomats refined successful experiences moving from one negotiation to the next.

In some respects, reviewing the SDGs at this moment in time could be considered premature as states have two-thirds of the remaining time to accomplish these goals. Indeed, the steep decline in the global economy due to the COVID–19 pandemic undoubtedly undermined the ability of all countries to better the lives of their citizens. Whether the global economy will recover in a timely fashion remains to be seen as stabilizing or increasing financial flows from North to South in all likelihood depends on having disposable income in the North.

Overall, the SDGs represent a significant improvement over the MDGs. The expansion of the goals and targets moved essential aspects of sustainability into the global limelight. The shift from a traditional neoliberal emphasis on economic development that attempted to build the South in the North's image gave way to a broader vision of sustainable development that also focused on social justice and the environment. However, this improvement does not mean that the SDGs do not have flaws.

Recalling from Chap. 11, the MDGs suffered from multiple criticisms. One crit- icism was that the technocratic initiation of one of the most comprehensive interna- tional agendas did not adequately consult states, especially those states from devel- oping countries [2]. Given that the technocrats working in the UN Secretariat drafted the MDGs, the OWG process should, in some sense, represent a return to normalcy in that states would assume control of the draft text. However, this configuration also represents an innovation from previous negotiating groups. In many intergov- ernmental negotiations, any member state that wanted to attend preliminary nego- tiations was historically welcomed. The creation of seats assigned to countries to share simultaneously limited the number of voices that could attend the meetings but enhanced the number of voices able to contribute to the negotiations. In other words, this innovation broke up the traditional UN triangle of negotiations between the United States, Europe, and the G77 and China. The so-called "troika" model effec- tively forced countries to negotiate directly with each other in a series of horizontal conversations rather than the pyramid structure [3].

For all the focus on the role of states in developing the SDGs, the reality is that the UN negotiation process is much more complicated. Not only was this complexity demonstrated during the negotiation process, but also by the result. The 2030 ASD created a complex data-driven system to achieve development based on neoliberal principles that incorporates a wide variety of actors, including the private sector, civil society, and the academic community. It thus requires a significant improvement in the capabilities of international organizations to carry out the tasks assigned by member states.

The UN also expanded its role as part of the vital international infrastructure for global negotiations. The hybrid negotiating structure utilized in creating the SDGs reaffirmed the state-centric negotiating focus, but it also increased the need for technocrats in the future. The significant expansion of goals, targets, and indicators to monitor the development status of the global South created and continues to create a need for an increasing number of technocrats to develop, analyze, and publicize the data produced. It also expanded the relationship between the UN and researchers as the data needs required technical expertise to create the new data sets.

While the UN has little control over the actions of civil society and the private sector, it has nevertheless found ways to encourage the cooperation of these groups. The creation of the PD4SDG database not only allows for capacity building for successful projects but also good publicity for their project proponents. It also subtly applies peer pressure to other civil society and private sector actors to participate by advertising the number of projects and actions taken. In doing so, the UN expanded its moral authority and legitimacy at a time in which domestic societies emphasized eliminating inequalities and achieving social justice. As a result, the UN continues to increase the number of entities willing to cooperate with the UN as its rhetoric emphasizes the global good.

A second criticism involved substance, namely, the MDGs' essential characteristics did not fully incorporate human rights and social justice into this program [13]. The SDGs attempted to correct for this deficiency by increasing the number of goals, targets, and indicators in an effort to include these issues and to create a sharp focus on each. Fukuda-Parr [7] reported that NGOs working in this area believed the SDGs to be superior to the MDGs in how the SDGs incorporated human rights into the document. Similarly, Gupta and Vegelin [9] argue that the SDGs fared well on incorporating "social inclusiveness." However, Gellers and Cheatham [8] note that the SDGs curiously omitted environmental justice from the goals and targets, although these authors argue that the concept of environmental justice (but not the words) was deeply embedded in the targets.

Scholarly efforts to critique the SDGs are not without their flaws. One of the primary premises of international relations theory, indeed, of most scholarly inquiry, is the search for a small number of theories or laws that work repeatedly. In the so-called "hard" sciences, scholars search to describe the physical world by establishing these laws. For example, the three laws of thermodynamics illustrate how energy moves from one object to another. In contrast, scholars consider social sciences to be a "soft" science based on the failure to discover similar laws that govern people's behaviors.

In the words of Ferguson and Mansbach [6], the elusive quest for one theory continues to consume considerable intellectual effort. There is one flaw with this elusive quest. Scholars assume that one primary law to describe all the human organization for the purpose of providing order exists. Stevens and Kanie [17] discussed one aspect of this approach when they observed that traditional analysis techniques focused on influence or effectiveness might yield "skepticism" instead of recognizing that UN diplomats successfully upgraded the UN paradigm.

There are two significant reasons why this search is doomed to fail. First, people are not constrained to act in the same manner as energy, chemicals, or biological processes. People have independent agency limited only by their own creativity in designing alternatives and by the actions of other independent actors around them. Expecting people to make decisions according to one specific model of how the international system works would require not only for scholars to agree as to what that one theory is, it would also require mass indoctrination of all other participants. Second, "Planet Earth" is itself a changing system. Humankind does not possess perfect information about how this planet works. Accordingly, the underlying risk and uncertainty stemming from this piecemeal knowledge also mean that at times and places, international environmental treaty making relies on glorified assumptions and guesswork to decide what is in the best interests of both people and the planet. As our scientific knowledge improves, public policy preferences also shift, causing actors to behave differently than they have in the past. This suggests that scholarly models produced by analyzing past behavior may be a poor predictor of future behavior.

A subtle shift occurred with the MDGs and expanded through the creation of the SDGs, in that UN diplomats no longer constrain themselves to consider one particular issue area at a time. The international environmental agenda no longer fits into neatly defined regimes, nor its successor, regime complexes. The creation and the implementation of the SDGs belie the likelihood that one simple scholarly model exists that can explain all international relations. International relations will, for the foreseeable future, remain an area of creativity and innovation that will at once frustrate and excite scholars, students, and diplomats for years to come.

References

1. Adams B (2015) SDG indicators and data: who collects? who reports? who benefits? Glob policy watch 9
2. Amin S (2006) The millennium development goals: a critique from the south. Mon Rev 57(10):1–15
3. Chasek P, Wagner LM (2016) Breaking the mold: a new type of multilateral sustainable development negotiation. Int Environ Agreem 16(3):397–413
4. Dimitrov RS (2020) Empty institutions in global environmental politics. Int Stud Rev 22(3):626–650
5. Egelston AE et al (2019) Networks for the future: a mathematical network analysis of the partnership data for sustainable development goals. Sustain 11(19):5511
6. Ferguson YH, Mansbach RW (2003) The elusive quest continues: theory and global politics. Pearson College Division, London, UK
7. Fukuda-Parr S (2016) From the millennium development goals to the sustainable development goals: shifts in purpose, concept, and politics of global goal setting for development. Gender Dev 24(1):43–52
8. Gellers JC, Cheatham TJ (2018) Sustainable development goals and environmental justice: realization through disaggregation. Wisc Int Law J 36:276

9. Gupta J, Vegelin C (2016) Sustainable development goals and inclusive development. Int Environ Agreem-P 16(3):433–448
10. Independent Expert Advisory Group on a Data Revolution for Sustainable Development (2014) A world that counts: mobilizing the data revolution for sustainable development. https://www.undatarevolution.org/wp-content/uploads/2014/11/A-World-That-Counts.pdf. Accessed 31 Jul 2022
11. Le Blanc D (2015) Towards integration at last? The sustainable development goals as a network of targets. Sustain Dev 23(3):176–187
12. Nilsson M et al (2016) Policy: map the interactions between sustainable development goals. Nat News 534(7607):320
13. Office of the High Commissioner for Human Rights (2008) Claiming the millennium development goals: a human rights approach. United Nations Publications
14. Pattberg P, Mert A (2013) The future we get might not be the future we want: analyzing the Rio + 20 outcomes. Glob Policy 4(3):305–310
15. Pearce F (2012) Beyond Rio, green economics can give us hope. The Guardian. http://www.guardian.co.uk/environment/2012/jun/28/rio-green-economics-hope
16. Pradhan P et al (2017) A systematic study of sustainable development goal (SDG) interactions. Earth's Future 5(11):1169–1179
17. Stevens C, Kanie N (2016) The transformative potential of the sustainable development goals (SDGs). Int Environ Agreem 16:393–396
18. UN DESA (2021a) SUST – RCNJ + HESI. https://sustainabledevelopment.un.org/partnership/?p=36692. Accessed 28 Feb 2021
19. UN DESA (2021b) 10% of marine protected areas. https://oceanconference.un.org/commitments/?id=20269. Accessed 28 Feb 2021
20. UN DESA (2021c) Tier classification for global SDG indicators. https://unstats.un.org/sdgs/iaeg-sdgs/tier-classification/. Accessed 3 Mar 2021
21. UN DESA (2021d) What we do. https://www.un.org/en/desa/what-we-do. Accessed 5 Mar 2021
22. UN GA (2010) Resolution adopted by the general assembly on 22 September 2010. A/RES/65/1
23. UN GA (2012) Resolution 66/288, the future we want. A/RES/66/288
24. UN System Task Team on the Post-2015 UN Development Agenda (2012). Realizing the future we want for all. Report to the Secretary-General. United Nations, New York
25. Weber H (2017) Politics of 'leaving no one behind': contesting the 2030 sustainable development goals agenda. Globalizations 14(3):399–414
26. Yiu LS, Saner R (2014) Sustainable development goals and millennium development goals: an analysis of the shaping and negotiation process. Asia Pac J Publ Adm 36(2):89–107

Chapter 15
Conclusions

Abstract International environmental diplomacy emerged in the 1970s out of a determination to protect planet Earth for present and future generations. While undoubtedly rooted in the geopolitical circumstances of the last half a century, international environmental diplomacy remains on the global agenda. Further, the UN adopted sustainable development as one of its primary missions. Progress, not perfection, remains an apt description for this realm of international diplomacy. States must not be allowed to ignore that the nature of international environmental problems requires cooperation as they cannot solve environmental issues alone. Supporters of environmental diplomacy have, in the past, been refreshingly optimistic that the next meeting will create the tipping point that brings people together to save the planet. Academic scholarship that continues to look at the present, but also strives to understand our past more fully can meaningfully contribute to global civic engagement that reminds all that this is a planet worth saving.

Keywords International environmental diplomacy · Stockholm + 50 · Governance · Sustainable development · UNCHE · Stockholm Action Plan

From the beginning of international environmental diplomacy's efforts to save this planet, diplomats and attendees of environmental conferences recognized the scale of the effort needed to halt environmental destruction and allow the Earth time to regenerate. This project attempted to provide a continuous story of these efforts. This field does not consist of episodic negotiating sessions. While meetings have a fixed start and end date, international environmental affairs constitute a continual process of change connected to the past while planning for a better future.

Upon receiving input from various important actors, diplomats embedded these goals within the Stockholm Action Plan in 1972. States noted that "Individuals in all walks of life as well as organizations in many fields, by their values and the sum of their actions, will shape the world environment of their future" [6: 3]. With the advantage of looking back at our efforts to save this planet, it is clear that these words ring truer today than at the time of their creation.

While the diplomats attending the UNCHE at Stockholm may not have envisioned the climate activism demonstrated by the Extinction Rebellion or Greta Thunberg's lecturing senior diplomats and heads of state, the global public cares greatly about

© The Author(s), under exclusive license to Springer Nature Switzerland AG 2022
A. Egelston, *Worth Saving*, AESS Interdisciplinary Environmental Studies and Sciences Series, https://doi.org/10.1007/978-3-031-06990-1_15

the state of the environment. This public concern elevated environmental norms into the popular culture and the everyday decision-making processes of individuals, corporations, and governments at all levels. Humankind widely acknowledges that decisions made collectively and individually impact the natural and built environment for better and for worse.

Building on the international diplomacy begun during the 1972 UNCHE, states branched out across a variety of issue areas that impact the quality of our environment to define a set of norms, values, behaviors, and actions by which global society acting in concert or as individuals may improve the quality of the environment. In creating these new environmental protections, global society joined together to articulate the concept of sustainable development. This paradigm seeks to reconcile the differences between the North and the South and do so in a manner that respects cultural differences and restores environmental equity.

The sheer number of individuals that engaged and continue to engage in this process brings great hope for the future. Many thousands of individuals participated in international environmental politics seeking to bridge the gaps in development between developed and developing countries and improve, if not end, environmental problems. This global civic engagement improved the quality of life for individuals at the local, national, and international levels. This great hope for further success is not unfounded. The quality of life for millions of people improved because of these efforts, present circumstances excluded from the COVID 19 pandemic. However, the work is, as of yet, unfinished.

This chapter turns to an evaluation of the entirety of the last fifty years of international environmental history to determine the status of international environmental diplomacy. Efforts to protect planet Earth neither consistently fail nor consistently succeed but vary between these two extremes. Incredible moments of collaboration improve environmental quality and increase states' willingness to collaborate in the future. Tragic collapses of international diplomacy allow environmental damage to continue while also prolonging exposures to increased environmental risks.

The first section reviews the original goals for international environmental affairs presented in the Stockholm Action Plan. Section two returns to the three characteristics of environmental affairs, complexity, change, and continuity. Section three presents concluding thoughts on scholarly activity that seeks to understand an expanding agenda. This book concludes by speculating about the future of international environmental diplomacy.

15.1 Does International Environmental Diplomacy Make a Difference?

At the UNCHE in 1972, international diplomats sought to protect the environment by creating new international norms, establishing a process for managing environmental diplomacy, and creating the outline of an ambitious environmental agenda.

While diplomats at Stockholm cleverly avoided determining deadlines for all these things, the Stockholm Declaration and Stockholm Action Plan nevertheless created an expectation that states adjust their behaviors. The UN system should add new infrastructure, and the recommended actions should occur as quickly as possible.

Certainly, planet Earth fared better because of the UNCHE meeting than it would have otherwise. A simple counterfactual argument would point out that UNCHE stimulated the creation of many environmental ministries (or equivalent) that halted environmental damage and led to consequences that punished miscreants and deterred future misbehaviors. A more complex counterfactual argument might well consider that a stronger environmental conference might have emerged in the aftermath of an environmental tragedy. These stronger environmental protections would take place around two equally important hypothetical conditions. First, planet Earth would continue to degrade in the time between the two conferences. Second, planet Earth and an unknown number of people would have suffered because of this hypothetical environmental tragedy.

Yet, it is also clear that portions of the UNCHE recommendations did not occur quickly. While the UN system attempted to accomplish all the recommendations, individual states routinely declined to cooperate. Perhaps the United States might quickly come to mind as the state least likely to ratify environmental treaties, and this is true. It is not the only country to decline to do so. Japan is notorious for its whaling; Brazil for its unwillingness to end deforestation of the Amazon rainforest.

Perhaps a better question may be whether international efforts to protect the environment made a difference in the overall health of the planet, its flora and fauna, and its residents. Measuring this outcome reliably often defies scientific research processes that require a transparent causal chain between diplomatic action and environmental improvement that is difficult, if not impossible, to establish. As a proxy, scholars assess whether treaties operate as planned or the international institution functions as designed. In this sense, international environmental diplomacy is neither a great failure nor an astounding success.

Thus, this manuscript turns to the more manageable question of the lasting impacts of the Stockholm Action Plan by examining its work processes, principles, and structures. First, the organization of international environmental diplomacy is composed of a three-step process of environmental assessment, environmental management, and supporting measures that form the blueprint for environmental diplomacy. The best scientists from every nation, with the most expertise on how the earth functions, have become deeply embedded in the admittedly political processes of healing the planet. Along with scientists, environmental policy specialists, health professionals, engineers, and a myriad of other professionals assist in providing information that states use to make decisions on behalf of their citizens.

Perhaps the work process that changed the most over time involves environmental management. While the language of the Stockholm Action Plan neglected to define this term, several possible interpretations of this role and function can reasonably be inferred from the context. In all likelihood, states intended to keep this role for themselves. However, NGOs disagreed with this assessment, believing instead that environmental management should be the responsibility of all to be effective. Thus, the ambiguity around this phrase proved helpful as environmental diplomacy has changed over time from a state-centric to a multi-centric approach with the rise of the SDGs as articulated in the 2030 ASD.

States have not lived up to expectations when reviewing supporting measures on behalf of environmental affairs. Financing for the environment frequently lags institutional needs. Equally importantly, environmental affairs require increasingly specialized knowledge that can only be acquired through lengthy and complex educational systems whose establishment remains out of reach for significant portions of developing countries, especially those in the LDCs. Efforts to build capacity, to transfer technology, and to share in the economic benefits of consuming natural resources, have not, to date, closed these gaps.

Second, the moral principles espoused at Stockholm have proven to be more mixed when evaluated through the lens of environmental protectionism. Existing customary international law conceptualizations such as do no harm and good neighborliness have proven difficult to achieve in light of environmental pollution, particularly air and water pollution, that easily defies state controls. Perhaps most disappointing is the hazardous waste trade, where chemicals with a known propensity to cause environmental damage are deliberately allowed to leave developed countries en route to developing countries that are less equipped to handle the complexities of containing the damage.

State sovereignty also has hampered international cooperation to stave off more significant environmental damage. States have been hesitant to curb chemical manufacturing or other industrial processes that damage the environment until ironclad proof emerges that the chemicals directly cause severe environmental damage. The United States, in particular, has been hesitant to ratify environmental treaties for a variety of reasons. The United States refuses to mandate widespread climate change reductions within the climate change regime until other prominent economies such as China also agree to make more substantial reductions. That is not to say that state sovereignty always impedes environmental cooperation or environmental protection. Interestingly, the Basel Convention helped protect the environment when states refused to allow these dangerous wastes and problematic chemicals into the country. Similarly, the Rotterdam Convention and Stockholm POPs Convention also utilized state sovereignty to protect the environment.

Newer principles such as the polluter pays, the precautionary principle, and common but differentiated responsibilities assisted with spreading environmentalism around the globe. States developed these three principles to encourage states to take more robust measures to protect the environment while leaving the current state system intact. While these principles undoubtedly aided states in protecting the environment, their importance pales when compared to the articulation of sustainable development. While sustainable development originated from environmental affairs, its reach is, in fact, much broader, crossing over from a principle to a paradigm.

Third, states created UNEP to ensure that global environmental affairs occupied a permanent, albeit niche place within international diplomacy. From this perspective, the last fifty years qualify as a resounding success. While UNEP's internal structure changed throughout this study, the organization historically played an intense and central role within international environmental governance. More than a few international treaties originated from the Governing Council, guiding UNEP's activities. As Ivanova [3] argues, UNEP remains a central organization within global environmental governance, regardless of whether its formal role in creating new international environmental treaties may be diminished.

Perhaps a more interesting question would be whether UNEP needs to have as vital a role today in this organization as it had in the 1970s, given the number of international regimes and regime complexes that have emerged in the last fifty years. Not only has the number of treaties increased, but the number of actors has also expanded, and the ease of communication, both formal and informal communications, multiplied. Today, there are more ways for issue areas to come to the forefront of the international environmental agenda. States, NGOs, the global media, and the global public have a unique ability to advance environmental protectionism.

If the need for a catalytic role to begin the processes of international environmental diplomacy decreased, the need for a professional organization to maintain the existing structure and expand the structure as needed increased. Keeping this intensive schedule of meetings, however, requires a professional, diplomatic staff with funding to not only run the conference but also to support the developing countries that otherwise would fail to attend. International environmental affairs could undoubtedly benefit from stable, long-term funding rather than relying on voluntary contributions or pledges that frequently do not materialize.

One of the fundamental questions within environmental affairs asks to what extent an organization is cost-efficient, bureaucratically well-managed, and effectively achieves its goals. Environmental organizations that failed to meet these criteria suffered severe criticism, reform, or in the worst cases, closure. Much, if not all, of the environmental bureaucracy experienced criticism and reform. UNEP itself was not immune to these changes as the 2012 reform movement replaced its Governing Council with the UN Environment Assembly in 2014.

New norms, regime complexes, regimes, and international legal principles have occurred as a result of the trajectory established at the UNCHE in 1972. More importantly, countries, corporations, and individuals changed their actions to improve and protect the environment that supports and surrounds life on planet Earth. The articulation of sustainable development during the 1980s advanced environmental protection by allowing greater cooperation and flexibility between the twin issues of environment and development.

The rise of sustainability that moved to the forefront of global environmental politics at the Earth Summit in 1992 remains the high point for international environmental policy. In the thirty years since this event, sustainable development served and continues to serve as the rallying cry for countries seeking a better life for their citizens. Perhaps more importantly, sustainable development proved able to move beyond the formal hallways of international diplomacy. NGOs and corporations alike found mechanisms that allowed them to contribute to environmental protectionism.

Governance through goals allowed environmental progress to move forward separate from the formal treaty-making processes. As such, this change represented a significant departure from past actions. In doing so, UN Secretary-General Annan increased the prospects for environmental improvements in the future. Thus, the MDG represented an essential innovation in international environmental diplomacy.

This innovation may enable the vital linkages between international diplomacy and environmental health to materialize. The expansion of the governance by goals represented by the 2030 ASD necessitated work on targets and indicators to assess environmental effectiveness. While this work remains ongoing, the 2030 ASD and the PD4SDG nudges entities to experiment on behalf of the environment. While some of these projects unquestionably failed to achieve their goals, others succeeded, and their success can be replicated to other areas of the world. This pattern of governance, while not as glamorous as the formal negotiating sessions, is no less important in terms of achieving environmentally friendly outcomes.

Whether the world retains the progress on environmental quality and development improvements in the face of the COVID-19 onslaught remains to be seen. What is known at the time of writing is that the world lost momentum toward achieving sustainable development [7]. The extent to which states will recommit to achieving these needed goals after the pandemic remains unknown.

Thus, the world's view turns to the upcoming Stockholm + 50 conference to be held in June 2022 in Stockholm, Sweden. International environmental diplomats and interested observers plan to gather to commemorate the UNCHE meeting that propelled environmental affairs into the spotlight. Expectations for this conference to fully reinvigorate rapid movement toward sustainable development barely exist. The conference duration is planned for a scant two days. Further, with the complexities of the COVID-19 pandemic limiting travel and constraining budgets, no major shifts in normative principles or concrete actions appear to be forthcoming. Instead, the Concept Note for the meeting emphasizes leadership dialogues along with time to reflect on the past as well as discuss how to move from commitment to action [8].

15.2 Complexity, Change, and Continuity Revisited

Measuring international environmental affairs ranges from simple metrics to systemwide changes that defy quantification. Simple metrics such as the number of countries and non-state actors involved in the negotiations, the number of treaties, the amount of media attention generated, or the number of protests that occurred in relation to the negotiating sessions may give a sense of the significant increases in the growth of this issue area over time. However, other changes in the international system come only by examining items beyond our ability to count or statistically analyze.

Over the course of this manuscript, complex dimensions of international affairs emerged, including the relationship between science and politics, the transnational nature of environmental pollution, differing state priorities between environmental protection and the environment, and differing models of how the world works that shaped analytical analysis of the international state system. This complexity leads to the conclusion that success in global efforts to enhance environmental protection lies both within and beyond the international negotiating sessions.

Significant action occurs in meetings away from these negotiating sessions that give shape and form to this field. International environmental affairs consist of formal negotiating sessions and a myriad of meetings between sessions that work on a wide variety of issues ranging from cooperation on scientific monitoring to harmonizing efforts in support of sustainable development across international organizations.

Environmental diplomats recognized the need to build change management systems into the work processes by which the UN system operates. One significant change in this system involves the shift in actors at the international level. At the beginning of international environmental affairs, the UN focused almost exclusively on member states. Today, the UN takes a broader view of its constituency as it attempts to influence states and non-states.

The international system managed this complexity by including change management systems into all three stages of its work processes: environmental assessment, environmental management, and supporting measures. While the UN initially attempted to build a global environmental monitoring system through Earthwatch, actual implementation fell short of this grand vision. Instead, individual issue areas designed their own mechanisms for interfacing with the scientific community. Environmental management expanded to incorporate the non-state actor as a technical advisor, political actor, and project implementer. The non-state actor also contributed to supporting measures by disseminating public information, providing education and training, and donating valuable goods and services to the UN.

Within the international negotiating sessions, diplomats recognized the need to incorporate change mechanisms into the environmental treaties by creating the framework convention-protocol system [5]. The advantages of this treaty system are many. First, building consensus on environmental principles has not been quick. Creating meaningful environmental treaties may take a decade or more. States do not typically adapt to changes in norms or patterns of behavior easily. Thus, this system allows states to consent to the agreeable and keep discussing the areas of discord. Second, the framework convention protocol allows states to take into account changes in both scientific knowledge and industrial technology.

While international diplomacy has changed dramatically over the last fifty years, the patterns of engagement nevertheless provide continuity. Ironically, the norms of environmental consciousness may provide more continuity than the contents or structures of the treaties themselves. The norms surrounding environmental affairs have been surprisingly consistent given the amount of change within individual issue areas. From the beginning of environmental affairs in the 1970s, normative principles such as the common heritage of humankind, additionality, common but differentiated responsibilities, and the precautionary principle continue to be deeply embedded within the international community. While perhaps ignored at the time of its creation, the Stockholm Declaration on the Environment contained within it the major strands of thought that would become sustainable development [2].

Perhaps the most important theme of this book relates to the many people who provided continuity to the international system through their consistent civic engagement. Individuals such as Maurice Strong and Dr. Mustafa Tolba provided leadership that created environmental norms, structures, and patterns of behavior that remain entrenched in the field. This field, however, also owes a debt of gratitude to lesser-known figures such as Rachel Carson, Barbara Ward, Julian Huxley, Peter Thatcher, or Thomas Lovejoy. These people sacrificed and toiled on behalf of the environment, riding the waves of intense public criticism and eventual public recognition.

15.3 The Future of Scholarship

This manuscript also highlighted excerpts of the academic literature from various subfields. While this text initially sought to organize and understand international environmental relations theory, the breadth of the events under consideration necessitated broadening explanatory factors into a dizzying array of subject areas, including political science, international relations, environmental studies, sociology, international law, and criminal justice perspectives. Further, this research occurred from various perspectives within these diverse fields. Consequently, analysts of international environmental politics may be well advised to consider events using models from multiple academic disciplines rather than adopting a one-size-fits-all approach.

The international system has not been shy about engaging individual scholars. Notable contributions from academia include Drs. Barbara Ward and Margaret Mead who lent credibility and prestige to the NGO community at a critical time during our initial global infrastructure in 1972. Similarly, Dr. Thomas Lovejoy used his expertise on conservation biology to impact the CBD in the 1980s and 1990s while Dr. Jeffrey Sachs lent his political acumen to the articulation of the MDGs as an advisor to Secretary-General Annan. In addition to these admittedly high profile researchers and consultants, lesser-known scholars participated through advising individual states, creating NGOs, and educating students who themselves will become the next generation of activists, business leaders, specialist advisors, and government diplomats. UN insiders enrich academic scholarship by sharing their insights with a broader audience.

Scholarly inquiry does not always happen contemporaneously with the time frame of the field under investigation. Thus, an attempt to structure contemporary scholarship chronologically failed for this reason. While this may frustrate students of international environmental politics beginning their journeys, this simple fact means that past events may well yield new insights into environmental politics. Further, a time gap between the event and the research activity does not diminish the importance of the insights and conclusions identified from the endeavor. However, the time factor may make research more difficult to achieve.

One of the key takeaways of this research is the need for scholars in all fields to contribute to understanding the processes that create international organizations. While the delineation between political science and history may be vague, the fiftieth anniversary may encourage environmental historians to focus on this field. Detailed work by many dedicated students of environmental diplomatic history will be necessary to fill in the gaps in our knowledge as the participants in these critical events may no longer be able to tell their individual and collective stories.

In the process of creating this manuscript, additional research opportunities accumulated rapidly. The role of non-state actors before the Earth Summit appears to be an area of future academic exploration. While scholarship focused on the role of states in creating international treaties, scholars did not evaluate non-state actors' participation in these events. Tantalizing hints of NGO activities suggest that NGOs' role as advocates for more stringent environmental protection began in the 1970s and continued steadily, even in the face of procedural handicaps created by the accreditation rules in place at that time.

Additionally, scholarship tends to focus on the outcomes of negotiating sessions, with noticeably less attention paid to the political processes that make these sessions possible. Traditional academic wisdom labels the 1980s as an almost lost cause, with a noticeable absence of international treaty negotiating sessions between the end of the UNCLOS III treaty process and the beginning of the climate change negotiations in 1988. The lack of formal negotiating sessions did not represent an end or a decline in thought on how best to protect the environment.

Reexamining this period may be worthwhile. Two critical pieces of infrastructure deserve more thorough exploration. The Montevideo Law Program and the UN Governing sessions created an international agenda that was successfully funded and executed in the late 1980s and 1990s. These meetings impacted the future of environmental protectionism in that they assist in determining the environmental agenda for the future. Additionally, the UN Environmental Assembly and Governing Council session prioritizes work that determines UNEP's priorities and assigns the budget to carry these actions out. States themselves recognized the importance of this body when they moved to a system of universal membership in 2012.

The long-term impact of the Earth Summit, including the climate change negotiations and the CBD, should also be reassessed as its major outcomes and new organizations did not meaningfully contribute to improvements in environmental quality or environmental management within the UN system. Agenda 21 did not penetrate global consciousness. The CSD, a body intended to oversee the implementation of the Earth Summit outcomes, ended in 2013. Dramatic impacts of climate change have already begun to appear. However, stabilizing the amount of greenhouse gases in the atmosphere does not appear to be on the horizon. During the writing of this manuscript, COP 26 took place in Glasgow, Scotland. This meeting concluded with no new meaningful reductions by major emitters, China and the United States.

While the CBD typically fared better than the climate change negotiations, the United States' decision to refrain from ratifying this treaty weakens its implementation. This weakness appears in its key protocols, the Cartagena Protocol and the Nagoya Protocol. Key developed countries such as Canada, Australia, and the Russian Federation also refrained from ratifying these agreements.

15.4   Hope for the Future

That is not to say that there is no hope for the future. Progress, not perfection, remains an apt description for this realm of international diplomacy. The international state system has proven to be resilient over the last four centuries, and it is unlikely that environmental problems will derail this system. However, there should be significant concern about the long-term impacts of climate change. States must not be allowed to ignore that the nature of international environmental problems requires cooperation as they cannot solve environmental issues alone. However, that does not mean that states should keep these problems around to motivate cooperation.

International environmental diplomacy worked in the past to improve the environment. The ozone hole continues to slowly close over the Arctic and the Antarctic, despite the unusually large ozone hole in 2020. Hazardous waste disposals are more likely to occur in an environmentally sound manner. Developing states successfully leveraged their sovereignty to ban the toxic trade for polluting more countries. Around the world, individual organizations continue to work to strengthen domestic environmental protections while corporations grow more mindful of their energy usage that spurs the production of greenhouse gases.

As the protests at the recent COP-26 meeting in Glasgow, Scotland, demonstrated, the global public has become increasingly educated about environmental problems. Public education and outreach efforts by NGOs and the UN system reach greater numbers of citizens through new data visualization techniques and improved access to social media outlets.

This education resulted in increased political mobilization to force states to act more appropriately to solve the pressing issues of climate change and environmental racism. Perhaps more encouragingly, young people around the world show signs of being more environmentally conscious than the generations that preceded them. Conceptualizations such as sustainability, with its focus on environmental protection and social justice, resonate with a generation interested in correcting the inequalities of the past.

The increase in the number of individuals interested in solving environmental problems brings with them increased energy for political action as well as considerable compassion and creativity in problem-solving. This creativity will not be limited to designing new political situations. Human ingenuity also creates new technologies that may decrease the environmental impacts as developed countries transition away from a carbon-intensive economy.

These reasons for hope do not negate the current global problems. As the COVID-19 pandemic enters its third year, the outlook for social justice globally has declined. Access to vaccines necessary to protect against the virus remains primarily limited to the Northern countries. Exposure to environmental risks continues to correspond to wealth both within and between countries. The number of people living in poverty rose and is anticipated to continue this rise as the pandemic continues. Likewise, the gap between the North and South, once thought to be narrowing slightly, reversed course and is widening.

Perhaps most telling in the quest to protect the planet is the wide variety of norms competing for attention within international environmental diplomacy. The dichotomy between environmental protection and economic wealth remains entrenched within some segments of a global society. It is difficult, if not impossible, to protect the planet when the countries charged with doing so cannot agree at either the domestic or international level what the problems are, much less what the solutions should be.

Given the clear conclusion that the international environmental system has, at best, partially achieved its goals of protecting people and the planet, what actions must we undertake in the future? Caldwell [1] stated that the primary rationale for locating environmental affairs at the international level stemmed from the reasoning that ecological damage occurred every way, every day. While this may be true, Caldwell's conclusion that the international state system should be the lead actor in solving this problem is unwarranted in many ways. Because environmental damage occurs everywhere, actions must be taken across all levels of government—local, state, regional, and international in order to protect the planet.

Certainly, there will remain a high profile role for international environmental diplomacy. However, individual contributions from high profile global elites and individual activists working toward a common goal to save planet Earth played a pivotal role in virtually every international environmental issue. Environmental affairs belong as much to the people as to governments or international organizations.

Meaningful individual civic engagement ranges from participating in organizations that continue to bring pressure on profit-driven corporations and environmentally recalcitrant states to providing citizen science to all levels of government. Environmental specialists in business and industry also have the potential to play pivotal roles in protecting planet Earth.

Koh [4] wrote that the creation of the UNCLOS III treaty reaffirmed the UN's collective commitment to cooperation for the purposes of protecting the environment. This commitment, while it has undoubtedly wavered over the course of the last half-century, nevertheless remains intact. International diplomacy continues to meet and work on the most pressing problems of humanity and resolve and refine the treaties of the past and present.

Supporters of environmental diplomacy have, in the past, been refreshingly optimistic that the next meeting will create the tipping point that brings people together to save the planet. This hope for the future is no less true today than yesterday. Despite the last half-century of struggle, there can truly be no doubt that international environmental diplomacy recognized a primary truth for all time, that this planet Earth, with its myriad forms of life, beautiful scenery, and abundant culture, is a planet worth saving.

References

1. Caldwell LK (1972) Defense of earth in a divided world. J Environ Health 35(3):228–236
2. Handl G (2012) Declaration of the United Nations conference on the human environment (Stockholm declaration), 1972 and the Rio declaration on environment and development, 1992. UN Audiovisual Library of International Law, 11
3. Ivanova M (2021) The untold story of the world's leading environmental institution: UNEP at fifty. MIT Press, Cambridge
4. Koh TT (1983) Negotiating a new world order for the sea. Va J Int Law 24:761
5. Montgomery MA (1990) Traveling toxic trash: an analysis of the 1989 Basel convention. Fletcher Forum World Aff 14(2):313–326
6. UNCHE (1972) Report on the United Nations conference on the human environment. A/CONF.48/14/Rev.1
7. UN DESA (2021) The sustainable development goals report 2021. https://unstats.un.org/sdgs/report/2021/The-Sustainable-Development-Goals-Report-2021.pdf. Accessed 4 Jan 2022
8. UNEP (2021) "Stockholm+50: a healthy planet for the prosperity of all—our responsibility, our opportunity" thought piece towards a concept note for the international meeting, 2–3 June 2022. https://wedocs.unep.org/bitstream/handle/20.500.11822/36939/STKLM50_HP.pdf. Accessed Jan 30 2022

Index

© The Editor(s) (if applicable) and The Author(s), under exclusive license to Springer Nature Switzerland AG 2022
A. Egelston, *Worth Saving*, AESS Interdisciplinary Environmental Studies and Sciences Series, https://doi.org/10.1007/978-3-031-06990-1

Printed in the United States
by Baker & Taylor Publisher Services